Praise for

Rory Sutherland and *Alchemy*

"*Alchemy*, as Rory Sutherland's title promised, turns words into gold. Veins of wisdom emerge regularly and brilliantly from these pages. Don't miss this book."

—Robert B. Cialdini, author of the million-copy bestseller *Influence*

"Sutherland, the legendary Vice Chairman of Ogilvy, uses his decades of experience to dissect human spending behavior in an insanely entertaining way. *Alchemy* combines scientific research with hilarious stories and case studies of campaigns for AmEx, Microsoft and the like. This is a must-read."

—*Entrepreneur*, "Best Books of the Year for Entrepreneurs"

"So many of us are trained to focus on data and the logical, and *Alchemy* makes a great argument for the irrational. I think everyone could use a reminder that asking dumb questions, reframing old ideas, and, in turn, trying to create a bit of magic, can lead to unexpected solutions for some of our most difficult problems. *Alchemy* was a reminder of that and then some."

—*Inc.*, "Great Books for Anyone Who Wants to Get Ahead in Life"

"The funniest book on [consumer behavior] is by Ogilvy Vice Chair Rory Sutherland. *Alchemy* is both a book on human behavior and a rallying cry to stand up against the spreadsheet mafia dominating most government and corporate policies today. . . . *Alchemy* is full of examples of how human behavior runs contrary to the laws of economics from Sutherland's work at Ogilvy."

—*Forbes*, "Best Books on Consumer Behavior to Help You in Business and in Life"

"Buy this book. I loved it. It's full of great insights."

—Matt Ridley, bestselling author of *The Rational Optimist* and *Genome*

"This is a breakthrough book: reality yields her secrets to practitioners, almost never to academics—something psychologists and economists, no matter what they say, are functionally unable to grasp. Rory Sutherland's *Alchemy* is the first such treatise written by someone who had true contact with reality via a P&L. Furthermore, it is funny as hell: I smiled and laughed at every paragraph. And *Alchemy* is wonderfully applicable to about everything in life, from how to announce airplane delays to how to handle unsold opera tickets. Buy two copies in case one is stolen."

—Nassim Nicholas Taleb, author of *The Black Swan*

"BRILLIANT, BRILLIANT, BRILLIANT: wonderfully heretical, naughty, funny, and wise. *Alchemy* has furnished me with enough good stories and sage aphorisms to keep my lecturers to tired students going for another ten years."

—Jules Goddard, London Business School

"Rory Sutherland's *Alchemy* is a deeply original book."

—Robert Trivers

"Sutherland is revelatory—and entertaining—in his anatomy of apparently paradoxical consumer choice in action."

—*The Times* (London)

"Sutherland brings together complex theory with stories and plentiful humour to both captivate and entertain. It's an important work for our era of apparent irrationality."

—*The Spectator* (UK)

"One of the leading minds in the world of branding."

—NPR's *TED Radio Hour*

"A hugely entertaining romp through ideas big and small."

—*The Irish Times*

"Will inspire you to think more ingeniously on your brands."

—Business Insider India

ALCHEMY

ALCHEMY: THE DARK ART AND CURIOUS SCIENCE OF CREATING MAGIC IN BRANDS, BUSINESS, AND LIFE

RORY SUTHERLAND

ch.
CUSTOM HOUSE

HarperCollins books may be purchased for educational, business, or sales promotional use. For information, please email the Special Markets Department at SPsales@harpercollins.com.

Originally published as *Alchemy: The Surprising Power of Ideas That Don't Make Sense* in the UK by WH Allen in 2019.

A hardcover edition of this book was published in 2019 by William Morrow, an imprint of HarperCollins Publishers.

FIRST CUSTOM HOUSE PAPERBACK EDITION PUBLISHED 2021.

Library of Congress Cataloging-in-Publication Data has been applied for.

ISBN 978-0-06-238842-1

24 25 26 27 28 LBC 11 10 9 8 7

CONTENTS

RORY'S RULES OF ALCHEMY vii
PROLOGUE: CHALLENGING COCA-COLA ix
INTRODUCTION:
CRACKING THE (HUMAN) CODE 1

1: ON THE USES AND ABUSES OF REASON 49
2: AN ALCHEMIST'S TALE
(OR WHY MAGIC REALLY STILL EXISTS) 133
3: SIGNALLING 165
4: SUBCONSCIOUS HACKING:
SIGNALLING TO OURSELVES 211
5: SATISFICING 245
6: PSYCHOPHYSICS 271
7: HOW TO BE AN ALCHEMIST 317

CONCLUSION:
ON BEING A LITTLE LESS LOGICAL 343

ENDNOTES 359
LIST OF ILLUSTRATIONS 361

RORY'S RULES OF ALCHEMY

1. The opposite of a good idea can also be a good idea.
2. Don't design for average.
3. It doesn't pay to be logical if everyone else is being logical.
4. The nature of our attention affects the nature of our experience.
5. A flower is simply a weed with an advertising budget.
6. The problem with logic is that it kills off magic.
7. A good guess which stands up to observation is still science. So is a lucky accident.
8. Test counterintuitive things only because no one else will.
9. Solving problems using rationality is like playing golf with only one club.
10. Dare to be trivial.
11. If there were a logical answer, we would have found it.

PROLOGUE: CHALLENGING COCA-COLA

Imagine that you are sitting in the boardroom of a major global drinks company, charged with producing a new product that will rival the position of Coca-Cola as the world's second most popular cold non-alcoholic drink.*

What do you say? How would you respond? Well, the first thing I would say, unless I were in a particularly mischievous mood, is something like this: 'We need to produce a drink that tastes nicer than Coke, that costs less than Coke, and that comes in a really big bottle so people get great value for money.' What I'm fairly sure nobody would say is this: 'Hey, let's try marketing a really expensive drink, that comes in a tiny can ... and that tastes kind of disgusting.' Yet that is exactly what one company did. And by doing so they launched a soft drinks brand that would indeed go on to be a worthy rival to Coca-Cola: that drink was Red Bull.

When I say that Red Bull 'tastes kind of disgusting', this is not a subjective opinion.† No, that was the opinion of a wide cross-section of the public. Before Red Bull launched outside of Thailand, where it had originated, it's widely rumoured that the licensee approached

* After water.

† I drink rather a lot of the stuff myself.

a research agency to see what the international consumer reaction would be to the drink's taste; the agency, a specialist in researching the flavouring of carbonated drinks, had never seen a worse reaction to *any* proposed new product.

Normally in consumer trials of new drinks, unenthusiastic respondents might phrase their dislike diffidently: 'It's not really my thing'; 'It's slightly cloying'; 'It's more a drink for kids' – that kind of thing. In the case of Red Bull, the criticism was almost angry: 'I wouldn't drink this piss if you paid me to,' was one refrain. And yet no one can deny that the drink has been wildly successful – after all, profits from the six billion cans sold annually are sufficient to fund a Formula 1 team on the side.

THE CASE FOR MAGIC

There is a simple premise to this book: that while the modern world often turns its back on this kind of illogic, it is at times uniquely powerful. Alongside the inarguably valuable products of science and logic, there are also hundreds of seemingly irrational solutions to human problems just waiting to be discovered, if only we dare to abandon standard-issue, naïve logic in the search for answers.

Unfortunately, because reductionist logic has proved so reliable in the physical sciences, we now believe it must be applicable everywhere – even in the much messier field of human affairs. The models that dominate all human decision-making today are duly heavy on simplistic logic, and light on magic – a spreadsheet leaves no room for miracles. But what if this approach is wrong? What if, in our quest to recreate the certainty of the laws of physics, we are now too eager to impose the same consistency and certainty in fields where it has no place?

Take work and holidays, for example. Some 68 per cent of Americans would *pay* to have two weeks more holiday than the meagre two weeks most enjoy at present – they would accept a 4 per cent pay cut in return for double the amount of vacation time.

But what if there were *no cost whatsoever* to increasing everyone's vacation allowance? What if we discovered that greater leisure time would benefit the US economy, both in terms of money spent on leisure goods and also in greater productivity? Perhaps people with more vacation time might be prepared to work for longer in life, rather than retiring to a Florida golf course as soon as it became affordable? Or perhaps they might simply be better at their jobs if they were reasonably rested and inspired by travel and leisure? Besides, it is now plausible that, for many jobs, recent advances in technology mean there is little difference in the contribution you make to your workplace, whether you are in a cubicle in Boise, Idaho or on a beach in Barbados.

There is an abundance of supporting evidence for these magical outcomes: the French are astonishingly productive on the rare occasion they are not on holiday; the German economy is successful, despite six weeks of annual leave being commonplace. But there is no model of the world that allows for America to contemplate, let alone trial, this possibly magical solution. In the left-brain, logical model of the world, productivity is proportional to hours worked, and a doubling of holiday time must lead to a corresponding 4 per cent fall in salary.

The technocratic mind models the economy as though it were a machine: if the machine is left idle for a greater amount of time, then it must be less valuable. But the economy is not a machine – it is a highly complex system. Machines don't allow for magic, but complex systems do.

Engineering doesn't allow for magic. Psychology does.

In our addiction to naïve logic, we have created a magic-free world of neat economic models, business case studies and narrow technological ideas, which together give us a wonderfully reassuring sense of mastery over a complex world. Often these models are useful, but sometimes they are inaccurate or misleading. And occasionally they are highly dangerous.

We should never forget that our need for logic and certainty brings costs as well as benefits. The need to appear scientific in our methodology may prevent us from considering other, less logical and more magical solutions, which can be cheap, fast-acting and effective. The mythical 'butterfly effect' does exist, but we don't spend enough time butterfly hunting. Here are some recent butterfly effect discoveries, from my own experience:

1. A website adds a single extra option to its checkout procedure – and increases sales by $300m per year.
2. An airline changes the way in which flights are presented – and sells £8m more of premium seating per year.
3. A software company makes a seemingly inconsequential change to call-centre procedure – and retains business worth several million pounds.
4. A publisher adds four trivial words to a call-centre script – and doubles the rate of conversion to sales.
5. A fast-food outlet increases sales of a product by putting the price ... up.

All these disproportionate successes were, to an economist, entirely illogical. All of them worked. And all of them, apart from the first, were produced by a division of my advertising agency, Ogilvy, which I founded to look for counter-intuitive solutions to problems. We discovered that problems almost always have a plethora of seemingly irrational solutions waiting to be discovered, but that nobody is looking for them; everyone is too preoccupied with logic to look anywhere else. We also found, rather annoyingly, that the success of this approach did not always guarantee repeat business; it is difficult for a company, or indeed a government, to request a budget for the pursuit of such magical solutions, because a business case has to look logical.

It's true that logic is usually the best way to succeed in an argument, but if you want to succeed in life it is not necessarily all that useful; entrepreneurs are disproportionately valuable precisely because they are *not* confined to doing only those things that make sense to a committee. Interestingly, the likes of Steve Jobs, James Dyson, Elon Musk and Peter Thiel often seem certifiably bonkers; Henry Ford famously despised accountants – the Ford Motor Company was never audited while he had control of it.

When you demand logic, you pay a hidden price: you destroy magic. And the modern world, oversupplied as it is with economists, technocrats, managers, analysts, spreadsheet-tweakers and algorithm designers, is becoming a more and more difficult place to practise magic – or even to experiment with it. In what follows, I hope to remind everyone that magic should have a place in our lives – it is never too late to discover your inner alchemist.

ALCHEMY

INTRODUCTION: CRACKING THE (HUMAN) CODE

I am writing this book with two screens in front of me, one of which is displaying a series of recent results from a test that my colleagues have just performed to try to increase the effectiveness of charity fundraising.

Once a year, volunteers for our client charity drop printed envelopes through millions of doors, and return a few weeks later to collect people's donations. This year the envelopes contained a hurricane relief appeal, but some of these envelopes were randomly different from the rest: 100,000 of them announced that the envelopes had been delivered by volunteers; 100,000 encouraged people to complete a form which meant their donation would be boosted by a 25 per cent tax rebate; 100,000 were in better-quality envelopes; and 100,000 were in portrait format (so the flap of the envelope was along the short side rather than the long one).

If you were an economist you would look at the results of this experiment and immediately conclude that people are completely insane. Logically, the only one of these changes that should affect whether people give is the one that reminds you that, for every £1 you donate, the government will give a further 25p. The other three tests are seemingly irrelevant; the paper quality, the orientation of the envelope and the fact that it was hand-delivered by a volunteer add nothing to the rational reasons to donate.

However, the results tell a different story. The 'rational' envelope in fact *reduces* donations by over 30 per cent compared to the plain control, while the other three tests increase donations by over 10 per cent. The higher-quality paper also attracts a significantly higher number of more significant donations of £100 or more. I hope that, by the time you finish reading this book, you might better understand why these crazy-sounding results may make a strange kind of sense.

The human mind does not run on logic any more than a horse runs on petrol.

What are the possible explanations for these results? Well, perhaps it feels more natural to put notes or cheques in an envelope with the flap on a shorter edge. Putting a cheque for £100 into a thick envelope feels more agreeable than putting it into one made of cheap paper. And a volunteer's effort of hand-delivering the envelope may prompt the urge to reciprocate: we appreciate the effort they have made. Perhaps the mention of a 25 per cent 'bonus' on their donation reduces the amount that people feel they need to give? Stranger still, it also reduced the proportion of people who gave anything at all; I'll be honest with you – I have no idea why this should be.

Here's the thing. To a logical person, there would have been no point in testing three of these variables, but they are the three that actually work. This is an important metaphor for the contents of this book: if we allow the world to be run by logical people, we will only discover logical things. But in real life, most things aren't logical – they are psycho-logical.

There are often two reasons behind people's behaviour: the ostensibly logical reason, and the real reason. I have worked in advertising and marketing for the last 30 years. I tell people I do it to make money, to build brands and to solve business problems; none of these are things I dislike, but, truthfully, I do it because I am nosy.

Modern consumerism is the best-funded social science experiment in the world, the Galapagos Islands of human weirdness. More important still, an ad agency is one of the few remaining safe spaces for weird or eccentric people in the worlds of business and government. In ad agencies, mercifully, maverick opinion is still broadly encouraged or at least tolerated. You can ask stupid questions or make silly suggestions – and still get promoted. This freedom is much more valuable than we realise, because to reach intelligent answers, you often need to ask really dumb questions.

In most corporate settings, if you suddenly asked 'Why do people clean their teeth?' you would be looked at as a lunatic, and quite possibly unsafe. There is after all an official, approved, logical reason why we clean our teeth: to preserve dental health and reduce cavities or decay. Move on. Nothing to see here. But, as I will explain later in this book, I don't think that's the real reason. For instance, if it is, why are 95 per cent of all toothpastes flavoured with mint?

Human behaviour is an enigma. Learn to crack the code.

My assertion is that large parts of human behaviour are like a cryptic crossword clue: there is always a plausible surface meaning, but there is also a deeper answer hidden beneath the surface.

5 *Across:* Does perhaps rush around (4)

To someone who is unfamiliar with cryptic crosswords it will seem almost insane that the correct answer to this clue is 'deer', because there is no hint of the animal in the surface meaning of the clue. A simple crossword would have a clue like 'Sylvan ruminants (4)'. But to a cryptic crossword aficionado, solving this clue is relatively simple – provided you accept that nothing is as it appears. The 'surface' of the clue has misled you to see 'does' and 'rush' as verbs, while both are actually nouns. 'Does' is here the plural of doe.* Rush is a reed. Reed 'around' – i.e. spelled backwards – is 'deer'.†

This insight is only possible once you know not to take the clue literally, and human behaviour is often cryptic in a similar sense;

---

* A deer, a female deer.

† The 'perhaps' is needed for purity, as not all deer are does – some are stags.

there is an ostensible, rational, self-declared reason why we do things, and there is also a cryptic or hidden purpose. Learning how to disentangle the literal from the lateral meaning is essential to solving cryptic crosswords, and it is also essential to understanding human behaviour.

To avoid stupid mistakes, learn to be slightly silly.

Most people spend their time at work trying to look intelligent, and for the last fifty years or more, people have tried to look intelligent by trying to look like scientists; if you ask someone to explain why something happened, they will generally give you a plausible-sounding answer that makes them seem intelligent, rational or scientific but that may or may not be the real answer. The problem here is that real life is not a conventional science – the tools which work so well when designing a Boeing 787, say, will not work so well when designing a customer experience or a tax programme. People are not nearly as pliable or predictable as carbon fibre or metal alloys, and we should not pretend that they are.

Adam Smith, the father of economics, identified this problem in the late eighteenth century,‡ but it is a lesson which many economists have been ignoring ever since. If you want to look like a scientist, it pays to cultivate an air of certainty, but the problem with attachment to certainty is that it causes people completely to misrepresent the nature of the problem being examined, as if it were a simple physics problem rather than a psychological one. There is hence an ever-present temptation to pretend things are more 'logical' than they really are.

‡ Indeed Ibn Khaldun, the father of sociology, perhaps saw it in the fourteenth century.

INTRODUCING PSYCHO-LOGIC

This book is intended as a provocation, and is only accidentally a work of philosophy. It is about how you and other humans make decisions, and why these decisions may differ from what might be considered 'rationality'. My word to describe the way we make decisions – to distinguish it from the artificial concepts of 'logic' and 'rationality' – is 'psycho-logic'. It often diverges dramatically from the kind of logic you'll have been taught in high school maths lessons or in Economics 101. Rather than being designed to be optimal, it has evolved to be useful.

Logic is what makes a successful engineer or mathematician, but psycho-logic is what has made us a successful breed of monkey, that has survived and flourished over time. This alternative logic emerges from a parallel operating system within the human mind, which often operates unconsciously, and is far more powerful and pervasive than you realise. Rather like gravity, it is a force that nobody noticed until someone put a name to it.

I have chosen psycho-logic as a neutral and non-judgemental term. I have done this for a reason. When we do put a name to non-rational behaviour, it is usually a word like 'emotion', which makes it sound like logic's evil twin. 'You're being emotional' is used as code for 'you're being an idiot'. If you went into most boardrooms and announced that you had rejected a merger on 'emotional grounds', you would likely be shown the door. Yet we experience emotions for a reason – often a good reason for which we don't have the words.

Robert Zion, the social psychologist, once described cognitive psychology as 'social psychology with all the interesting variables set to zero'. The point he was making is that humans are a deeply social species (which may mean that research into human behaviour or choices in artificial experiments where there is no social context isn't really all that useful). In the real world, social context is absolutely critical. For instance, as the anthropologist Pierre Bourdieu observes, gift giving is viewed as a good thing in most human societies, but it only takes a very small change in context to make a gift an insult rather than a blessing; returning a present to the person who has given it to you, for example, is one of the rudest things you can do. Similarly, offering people money when they do something you like makes perfect sense according to economic theory and is called an incentive, but this does not mean you should try to pay your spouse for sex.*

The alchemy of this book's title is the science of knowing what economists are wrong about. The trick to being an alchemist lies not in understanding universal laws, but in spotting the many instances where those laws do not apply. It lies not in narrow logic, but in the equally important skill of knowing when and how to abandon it. This is why alchemy is more valuable today than ever.

---

* As an experiment, I tried this once – about three months later, I was offered some sex. So the economic approach, if it works at all, works rather slowly.

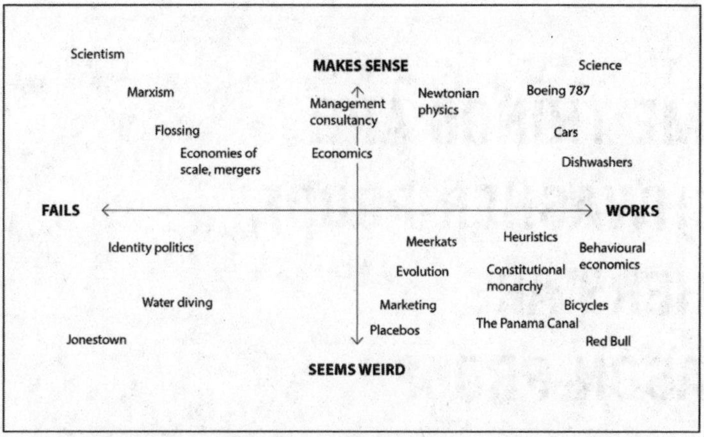

Not everything that makes sense works, and not everything that works makes sense. The top-right section of this graph is populated with the very real and significant advances made in pure science, where achievements can be made by improving on human perception and psychology. In the other quadrants, 'wonky' human perception and emotionality are integral to any workable solution.

The bicycle may seem a strange inclusion here: however, although humans can learn how to ride bicycles quite easily, physicists still cannot fully understand how bicycles work. Seriously. The bicycle evolved by trial and error more than by intentional design.

SOME THINGS ARE DISHWASHER-PROOF, OTHERS ARE REASON-PROOF

Here's a simple (if expensive) lifestyle hack. If you would like everything in your kitchen to be dishwasher-proof, simply *treat* everything in your kitchen as though it was; after a year or so, anything that isn't dishwasher-proof will have been either destroyed or rendered unusable. Bingo – everything you have left will now be dishwasher-proof! Think of it as a kind of kitchen-utensil Darwinism.

Similarly, if you expose every one of the world's problems to ostensibly logical solutions, those that can easily be solved by logic will rapidly disappear, and all that will be left are the ones that are logic-proof – those where, for whatever reason, the logical answer does not work. Most political, business, foreign policy and, I strongly suspect, marital problems seem to be of this type.

This isn't the Middle Ages, which had too many alchemists and not enough scientists. Now it's the other way around; people who are very good at deploying and displaying conventional, deductive logic are everywhere, and they're usually busily engaged in trying to apply some theory or model to something in order to optimise it. Much of the time, this is a good thing. I don't want a conceptual artist in charge of air-traffic control, for instance. However, we now unfortunately fetishise logic to such an extent that we are increasingly blind to its failings.

For instance, the victorious Brexit campaign in Britain and the election of Donald Trump in the United States have both been routinely blamed on the clueless and emotional behaviour of undereducated voters, but you could make equally strong cases that the Remain campaign in Britain and Hillary Clinton's failed bid for the American presidency failed because of the clueless, hyperrational behaviour of overeducated advisors, who threw away huge natural advantages. At one point we in Britain were even warned that 'a vote to leave the EU might result in rising labour costs' – by a highly astute businessman* who was so enraptured with models of economic efficiency that he was clearly unaware most voters would understand a 'rise in labour costs' as meaning a 'pay rise'.

Perhaps most startlingly of all, every single one of the Remain campaign's arguments resorted to economic logic, yet the EU is patently a political project, which served to make them seem greedy rather than principled, especially as the most vocal Remain supporters came from a class of people who had done very nicely out of globalisation. Notice that Winston Churchill did not urge us to fight the Second World War 'in order to regain access to key export markets'.

More data leads to better decisions. Except when it doesn't.

Across the Atlantic, meanwhile, the Clinton campaign was dominated by a strategist called Robby Mook, who had become so enamoured of data and mathematical modelling that he refused to use anything else. He derided Bill Clinton for suggesting he should connect the campaign with white working-class voters in the Midwest, mimicking a 'Grampa Simpson' voice to mock the former president† and dismissing another suggestion with the smug 'my data disagree with your anecdotes'.

---

* Stuart Rose, former executive chairman of Marks & Spencer.

† Whatever else you may think of Bill Clinton, his track record clearly indicates that he is an instinctive political genius.

Yet perhaps the anecdotal evidence was right, because the data was clearly wrong. Clinton did not visit Wisconsin once in the entire campaign, wrongly assuming that she would win there easily. Some in her team suggested that she should visit in the last days before the election, but the data told her to go to Arizona instead. Now I'm British, and have only been to Arizona four or five times, and Wisconsin twice. But even I would have said, 'that decision sounds weird to me'. After all, nothing I have ever seen in Wisconsin suggested that it was a state that would never vote for Donald Trump, and it has always had a strong streak of political eccentricity.

The need to rely on data can also blind you to important facts that lie outside your model. It was surely relevant that Trump was filling sports halls wherever he campaigned, while Clinton was drawing sparse crowds. It's important to remember that big data all comes from the same place – the past. A new campaigning style, a single rogue variable or a 'black swan' event can throw the most perfectly calibrated model into chaos. However, the losing sides in both these campaigns have never once considered that their reliance on logic might been the cause of their defeats, and the blame was pinned on anyone from 'Russians' to 'Facebook'. Maybe they were blameworthy in part, but no one has spent enough time asking whether an overreliance on mathematical models of decision-making might be to blame for the fact that in each case the clear favourite blew it.

In theory, you can't be too logical, but in practice, you can. Yet we never seem to believe that it is possible for logical solutions to fail. After all, if it makes sense, how can it possibly be wrong?

To solve logic-proof problems requires intelligent, logical people to admit the possibility that they might be wrong about something, but these people's minds are often most resistant to change – perhaps because their status is deeply entwined with their capacity for reason. Highly educated people don't merely *use* logic; it is part of their identity. When I told one economist that you can often increase the sales of a product by increasing its price, the reaction was one not of curiosity but of anger. It was as though I had insulted his dog or his favourite football team.

Imagine if it were impossible to get a well-paid job, or to hold political office, unless you supported the New York Yankees or Chelsea Football Club. We would regard such partisanship as absurd, yet devoted fans of logic control the levers of power everywhere. The Nobel Prize-winning behavioural scientist Richard Thaler said, 'As a general rule the US Government is run by lawyers who occasionally take advice from economists. Others interested in helping the lawyers out need not apply.'

Today it sometimes seems impossible to get a job without first demonstrating that you are in thrall to logic. We flatter such people through our education system, we promote them to positions of power and are subjected every day to their opinions in the newspapers. Our business consultants, accountants, policy-makers and think-tank pundits are all selected and rewarded for their ability to display impressive flights of reason.

This book is not an attack on the many healthy uses of logic or reason, but it is an attack on a dangerous kind of logical overreach, which demands that every solution should have a convincing rationale before it can even be considered or attempted. If this book provides you with nothing else, I hope it gives you permission to suggest slightly silly things from time to time. To fail a little more often. To think *unlike* an economist. There are *many* problems which are logic-proof, and which will never be solved by the kind of people who aspire to go to the World Economic Forum at Davos.‡ Remember the story of those envelopes.

We could never have evolved to be rational – it makes you weak.

‡ A bizarre international junket where, for some reason, the world's most intelligent people collectively decide that it is a good idea to spend part of January halfway up a mountain.

Now, as reasonable people, you're going to hate me saying this, and I don't feel good saying it myself. But, for all the man's faults, I think Donald Trump can solve many problems that the more rational Hillary Clinton simply wouldn't have been able to address. I don't admire him, but he is a decision maker from a different mould. For example, both candidates wanted manufacturing jobs to return to the United States. Hillary's solution was logical – engagement in tripartite trade negotiations with Mexico and Canada. But Donald simply said, 'We're going to build a wall, and the Mexicans are going to pay.'

'Ah,' you say. 'But he's never going to build that wall.' And I agree with you – I think it highly unlikely that a wall will be built, and even less likely that the unlucky Mexicans will agree to pay for it. But here's the thing: he may not need to build the wall to achieve his trade ambitions – he just needs people to believe that he might. Similarly, he doesn't need to repeal the North American Free Trade Agreement – he just needs to raise it as a possibility. Irrational people are much more powerful than rational people, because their threats are so much more convincing.

For perhaps thirty years, the prevailing economic consensus meant that no American carmaker felt they owed any patriotic duty to workers in their home country; had you suggested such a thing in any of their board meetings, you would have been viewed as a dinosaur. So pervasive was the belief in untrammelled free trade – on both sides of the American political divide – that manufacturing was shifted overseas without any consideration about whether there might be a risk to losing the support of government or public opinion. All Trump needed to do was to signal that this assumption was no longer safe. No tariffs (or walls) are actually needed: the threat of them alone is enough.§

A rational leader suggests changing course to avoid a storm. An irrational one can change the weather.

§ Hillary could not convincingly have made such a threat, because everyone would have known it was hollow. Trump is crazy enough to go through with it.

Being slightly bonkers can be a good negotiating strategy: being rational means you are predictable, and being predictable makes you weak. Hillary thinks like an economist, while Donald is a game theorist, and is able to achieve with one tweet what would take Clinton four years of congressional infighting. That's alchemy; you may hate it, but it works.

Some scientists believe that driverless cars will not work unless they learn to be irrational. If such cars stop reliably whenever a pedestrian appears in front of them, pedestrian crossings will be unnecessary and jaywalkers will be able to march into the road, forcing the driverless car to stop suddenly, at great discomfort to its occupants. To prevent this, driverless cars may have to learn to be 'angry', and to occasionally maliciously fail to stop in time and strike the pedestrian on the shins.

If you are wholly predictable, people learn to hack you.

CRIME, FICTION AND POST-RATIONALISM: OR WHY REALITY ISN'T NEARLY AS LOGICAL AS WE THINK

Think of life as like a criminal investigation: a beautifully linear and logical narrative when viewed in retrospect, but a fiendishly random, messy and wasteful process when experienced in real time. Crime fiction would be unreadably boring if it accurately depicted events, because the vast majority of it would involve enquiries that led nowhere. And that's how it's supposed to be – the single worst thing that can happen in a criminal investigation is for everyone involved to become fixated on the same theory, because one false assumption shared by everyone can undermine the entire investigation. There's a name for this – it's called 'privileging the hypothesis'.

A recent example of this phenomenon emerged during the bizarre trial of Amanda Knox and Raffaele Sollecito for the murder of Meredith Kercher in Perugia, Italy. It became impossible for the investigator and his team to see beyond their initial suspicion that, after Kercher had been killed, the perpetrator had staged a break-in to 'make it look like a burglary gone wrong'. Since no burglar from outside would need to stage a break-in, their only conclusion was that the staging took place to divert attention from the other flatmates and to disguise the fact that it was an inside job. Unfortunately, the initial suspicion was incorrect.

I sympathise a little with their attachment to the theory. After all, the break-in did, at first glance, look as though it might have been faked: there was some broken glass *outside* the window and an absence of footprints. But the theory of an inside job staged to look like a botched burglary was so doggedly held that all subsequent contradictory evidence was either suppressed or not shared with the press, and the result was a nonsense.

The break-in did look rather absurd at first glance – why would you break into a flat from a relatively exposed upstairs window? – until you realise that the purpose of breaking a window was not to gain access to the house, but to make a hell of a lot of noise while standing in a place from which an easy escape was possible. It thus helped the perpetrator ascertain with some confidence that there was no one around; if you smash a window and nobody intervenes, you can be fairly sure no one is going to notice you climbing through the same window five minutes later, but if a light goes on and a dog starts barking, you can simply leg it.

This example goes to the heart of how we see the world. Do we look at things from a single perspective, where you do one thing to achieve another, or do we accept that complex things are rather different? In a designed system, such as a machine, one thing does serve one narrow purpose, but in an evolved or complex system, or in human behaviour, things can have multiple uses depending on the context within which they are viewed.

The human mouth allows you to eat, but if your nose is blocked, it also allows you to breathe. In a similar way, it seems illogical to break into a building using the noisiest means possible, until you understand the context in which the offender is operating. It is not appropriate to bring the same habits of thought that we use to deal with things that have been consciously designed to understanding complex and evolved systems, with second-order considerations.

My problem with Marxism is that it makes too much sense.

THE DANGER OF TECHNOCRATIC ELITES

If you are a technocrat, you'll generally have achieved your status by explaining things in reverse; the plausible post-rationalisation is the stock-in-trade of the commentariat. Unfortunately, it is difficult for such people to avoid the trap of assuming that the same skills that can explain the past can be used to predict the future. Like a criminal investigation, what looks neat and logical when viewed with hindsight is usually much messier in real time. The same is true of scientific progress. It is easy to depict a discovery, once made, as resulting from a logical, and linear process, but that does not mean that science should progress according to neat, linear and sequential rules.

There are two separate forms of scientific enquiry – the discovery of what works and the explanation and understanding of *why* it works. These are two entirely different things, and can happen in either order. Scientific progress is not a one-way street. Aspirin, for instance, was known to work as an analgesic for decades before anyone knew *how* it worked. It was a discovery made by experience and only much later was it explained. If science did not allow for

* Bakelite, penicillin, the microwave, X-rays, radar, radio were all discovered 'backwards'.

such lucky accidents,* its record would be much poorer – imagine if we forbade the use of penicillin, because its discovery was not predicted in advance? Yet policy and business decisions are overwhelmingly based on a 'reason first, discovery later' methodology, which seems wasteful in the extreme. Remember the bicycle.

Evolution, too, is a haphazard process that discovers what can survive in a world where some things are predictable but others aren't. It works because each gene reaps the rewards and costs from its lucky or unlucky mistakes, but it doesn't care a damn about reasons. It isn't necessary for anything to make sense: if it works it survives and proliferates; if it doesn't, it diminishes and dies. It doesn't need to know *why* it works – it just needs to work.

Perhaps a plausible 'why' should not be a pre-requisite in deciding a 'what', and the things we try should not be confined to those things whose future success we can most easily explain in retrospect. The record of science in some ways casts doubt on a scientific approach to problem solving.

ON NONSENSE AND NON-SENSE

I'll admit it: I have only become qualified to write this book by accident. I am a classicist, not an anthropologist, but have, almost by chance, spent 30 years in the advertising industry – mostly in what is known as 'direct response', the form of advertising where people are urged to respond directly to your advertisement. It consists of well-funded behavioural experiments on a grand scale, and what this teaches us is that the models of human behaviour devised and promoted by economists and other conventionally rational people are wholly inadequate at predicting human behaviour.

What are the great achievements of economics? Ricardo's Theory of Comparative Advantage, perhaps? Or *The General Theory of Employment, Interest and Money* by John Maynard Keynes? And what is the single most important finding of the advertising industry? Perhaps it is that 'advertisements featuring cute animals tend to be more successful than ads that don't'.

I'm not joking. I recently had a meeting with a client where I learned that a customer prize draw to win 'free energy for a year – worth over £1,000' received 67,000 entries. The subsequent draw, where you could win a cute penguin nightlight (with a value of £15) received over 360,000 entries. One customer even turned down an offer of a £200 refund on their bill, saying, 'No, I'd rather have a penguin.' Even though I know this is true, so great is my desire to appear rational that I would find it very hard to stand in front of a board of directors and recommend that their advertising should feature rabbits, or perhaps a family of lemurs, because it sounds like nonsense. It isn't, though. It's a different kind of thing, which I call 'non-sense'.

Behavioural economics is an odd term. As Warren Buffett's business partner Charlie Munger once said, 'If economics isn't behavioural, I don't know what the hell is.' It's true: in a more sensible world, economics would be a subdiscipline of psychology.* Adam Smith was as much a behavioural economist as an economist – *The Wealth of Nations* (1776) doesn't contain a single equation. But, strange though it may seem, the study of economics has long been detached from how people behave in the real world, preferring to concern itself with a parallel universe in which people behave as economists *think they should*. It is to correct this circular logic that behavioural economics – made famous by experts such as Daniel Kahneman, Amos Tversky, Dan Ariely and Richard Thaler – has come to prominence. In many areas of policy and business there is much more value to be found in understanding how people behave in reality than how they should behave in theory.†

Behavioural economics might well be described as the study of the nonsensical and the non-sensical aspects of human behaviour. Sometimes our behaviour is nonsensical because we evolved for conditions different to those we now find ourselves in.‡ However, much 'irrational' human behaviour is not really nonsensical at all; it is non-sensical. For instance, viewed through the lens of

* The dissident Austrian School of economists wisely believed this.

† I know. Who would have thought it?

‡ For example, we probably love sugar too much: in the ancestral environment there was no refined sugar, and the only food with a comparable glycemic load was honey.

evolutionary psychology, the effectiveness of cute animals in advertising should not shock us. Advertising exists to be noticed, and we have evolved, surely, to pay attention to living things. An evolutionary psychologist might also suggest that a penguin nightlight – a gift for one's child – might be more emotionally rewarding than a cash reward, which is a gain for oneself.§

Sometimes human behaviour that seems nonsensical is really non-sensical – it only appears nonsensical because we are judging people's motivations, aims and intentions the wrong way. And sometimes behaviour is non-sensical because evolution is just smarter than we are. Evolution is like a brilliant uneducated craftsman: what it lacks in intellect it makes up for in experience.

For instance, for a long time the human appendix was thought to be nonsense, a vestigial remnant of some part of the digestive tract, which had served a useful purpose in our distant ancestors. It is certainly true that you can remove people's appendices and they seem to suffer no immediate ill effects. However, in 2007, William Parker, Randy Bollinger and their colleagues at Duke University in North Carolina hypothesised that the appendix actually serves as a haven for bacteria in the digestive system that are valuable both in aiding digestion and in providing immunity from disease. So, just as miners in the California Gold Rush would guard a live sourdough yeast 'starter' in a pouch around their necks, the body has its own pouch to preserve something valuable. Research later showed that individuals whose appendix had been removed were four times more likely to suffer from *clostridium difficile colitis*, an infection of the colon.

Given that cholera was a huge cause of death only a few generations ago, and given that it is thought by some to be making a

§ My friend, the evolutionary biologist Nichola Raihani, recently had her child's bicycle helmet stolen. She was immediately struck by the strength of her outrage, which was far more extreme than if her own bicycle helmet had been stolen.

comeback, perhaps the appendix should no longer be treated as disposable – it seems that, rather like the Spanish royal family, most of the time it's pointless or annoying, but sometimes it's invaluable.¶

Be careful before calling something nonsense.

The lesson we should learn from the appendix is that something can be valuable without necessarily being valuable all the time. Evolution does not take such a short-term, instrumentalist view. In looking for the *everyday* function of the human appendix, we were looking for the wrong thing. Whether something makes sense in theory matters less than whether it works in practice.

Like quite a few fellow Anglicans (but unlike my wife who is a priest and hospital chaplain) I am not quite sure of the existence of God, but I would be reluctant to disparage religion as nonsense, as some people do.

In a 1996 survey on the place of religion in public life in America, the Heritage Institute found that:

1. Churchgoers are more likely to be married, less likely to be divorced or single and more likely to manifest high levels of satisfaction in their marriage.
2. Church attendance is the most important predictor of marital stability and happiness.
3. The regular practice of religion helps poor people move out of poverty. Regular church attendance, for example, is

¶ Spain's peaceful and robust transition to democracy after Franco might have been impossible without the decisive role played by an arbitrary and symbolic head of state.

particularly instrumental in helping young people escape the poverty of inner-city life.

4. Regular religious practice generally inoculates individuals against a host of social problems, including suicide, drug abuse, out-of-wedlock births, crime and divorce.
5. The regular practice of religion also encourages such beneficial effects on mental health as less depression, higher self-esteem and greater family and marital happiness.
6. In repairing damage caused by alcoholism, drug addiction and marital breakdown, religious belief and practice are a major source of strength and recovery.**
7. Regular practice of religion is good for personal physical health: it increases longevity, improves one's chances of recovery from illness and lessens the incidence of many killer diseases.

Religion feels incompatible with modern life because it seems to involve delusional beliefs, but if the above results came from a trial of a new drug, we would want to add it to tap water. Just because we don't know why it works, we should not be blind to the fact that it does.††

Business, creativity and the arts are full of successful non-sense. In fact the single greatest strength of free markets is their ability to generate innovative things whose popularity makes no sense. Non-sense includes things that are useful or effective, even though (or perhaps because) they defy conventional logic.

Almost all good advertising contains some element of non-sense. At first glance this might make it look silly – it can certainly make selling it to a sceptical group of clients painfully embarrassing.

** Alcoholics Anonymous is, remember, modelled on explicitly religious principles.

†† Take that, Dawkins!

Imagine you are the board of an airline and have just spent three hours debating whether to buy 13 Airbus A350s or 11 Boeing 787s, each of which costs around $150 million. At the end of the meeting, you are presented with an idea for an advertising campaign that does not show an aircraft at all, but instead proposes to focus on the cucumber sandwiches and scones that might be served on board. This is non-sense – however, around 90 per cent of people have no idea what sort of aircraft they are travelling on or how a jet engine works but will infer a great deal about the safety and quality of the experience offered by an airline from the care and attention it pays to on-board snacks.‡‡

Presenting such things in a business setting packed with MBA graduates is slightly embarrassing; you start to envy people in IT or tax-planning, who can go into a meeting with rational proposals on a chart or spreadsheet. However, this fixation with sense-making can prove expensive. Imagine you are a company whose product is not selling well. Which of the following proposals would be easier to make in a board meeting called to resolve the problem? a) 'We should reduce the price' or b) 'We should feature more ducks in our advertising'. The first, of course – and yet the second could, in fact, be much more profitable.

This is a book written in defence of things that don't quite make sense, but it is also a book that – conversely – attacks our fetishisation of things that *do*. Once you accept that there may be a value or

‡‡ The gin brand Hendrick's engaged in a very clever bit of non-sense, when they suggested that their product be served not with lemon but with cucumber, which gained immediate salience. Being British, I failed to notice the genius of this move, which was that it also positioned the drink as sophisticatedly British in the United States; Americans find cucumber sandwiches a British peculiarity. To a Brit, of course, a cucumber is not seen as being particularly British – it is just something we make sandwiches with.

purpose to things that are hard to justify, you will naturally come to another conclusion: that it is perfectly possible to be both rational and wrong.

Logical ideas often fail because logic demands universally applicable laws but humans, unlike atoms, are not consistent enough in their behaviour for such laws to hold very broadly. For example, to the despair of utilitarians, we are not remotely consistent in whom we choose to help or cooperate with. Imagine that you get into financial trouble and ask a rich friend for a loan of £5,000, who patiently explains that you are a much less needy and deserving case for support than a village in Africa to which he plans to donate the same amount. Your friend is behaving perfectly rationally. Unfortunately he is no longer your friend.

It is impossible for human relations to work unless we accept that our obligations to some people will always exceed our obligations to others. Universal ideas like utilitarianism are logical, but seem not to function with the way we have evolved. Perhaps it is no coincidence that Jeremy Bentham, the father of utilitarianism, was one of the strangest and most anti-social people who ever lived.[§§]

The drive to be rational has led people to seek political and economic laws that are akin to the laws of physics – universally true and applicable. The caste of rational decision makers requires generalisable laws to allow them confidently to pronounce on matters without needing to consider the specifics of the

§§ It has often been proposed that he was autistic. I am reluctant to use this diagnosis too widely, but it is perhaps true that he was overburdened with the use of reason. He once declined the chance to meet his young nieces, saying, 'If I don't like them, I will not enjoy the experience, and if I do like them then I will be sad to see them leave.' Perfectly reasonable, I suppose, but weird as hell! Kant was also a weirdo.

situation.¶¶ And in reality 'context' is often the most important thing in determining how people think, behave and act: this simple fact dooms many universal models from the start.*** Because in order to form universal laws, naïve rationalists have to pretend that context doesn't matter.

¶¶ Notice that ordinary people are never allowed to pronounce on complex problems. When do you ever hear an immigration officer interviewed about immigration, or a street cop interviewed about crime? These people patently know far more about these issues than economists or sociologists, and yet we instead seek wisdom from people with models and theories rather than actual experience.

*** For instance, will wealthy Germans help poorer Germans? Yup. Will they help Syrians? Yes, albeit reluctantly. Poor Greeks, however? No chance.

THE OPPOSITE OF A GOOD IDEA CAN BE A GOOD IDEA

Economic theory is perhaps the most overambitious attempt to create universal rules of human behaviour – 'markets in everything', as the phrase goes. Yet it is all too common, in certain settings, for people's behaviour to run directly counter to the supposedly logical beliefs of standard economics. Take London housing, for example. Logic would suggest that, as house prices in London continue to rise, many Londoners who do not need to live in the city would decide to buy houses further away, gaining from price rises and relaxing the pressure on the market. In reality it seems the opposite happens: when sitting on a rising asset, people who would secretly prefer to move 50 or 200 miles away from London are reluctant to, for fear either that they will miss out on future price increases or that, once they leave, they will be unable to afford to move back again. Even though this is perfectly plausible – indeed it seems to be what often happens in reality – economics treats all markets as if they were the same. In the crude oil market, for instance, things might happen in line with economic predictions and rising prices may drive asset owners to sell, but markets for housing and oil are very different.

Does a tax rise cause you to work less because the returns for your labour are lower, or does it cause you to work harder, in order to maintain your present level of disposable wealth? It kind of depends. Logic requires that people find universal laws, but outside of scientific fields, there are fewer of these than we might expect. And once human psychology has a role to play, it is perfectly possible for behaviour to become entirely contradictory. For instance, there are two equally potent, but completely contradictory, ways to sell a product: 'Not many people own one of these, so it must be good' and 'Lots of people already own one of these, so it must be good.' As the brilliant Robert Cialdini highlights in *Influence: The Psychology of Persuasion*, the principles of selling and behaviour change are imbued with contradictions.

On the one hand, luxury goods would be destroyed if they were too widespread – no one would want a designer bag that was owned by five million other people.* On the other hand, many foodstuffs seem to be popular *only* because they are popular. I have always been puzzled by the popularity of miso soup. Imagine if it did not exist, but one day my daughter brought me a bowl of it: 'Look, Dad, I've just invented a new soup.' After removing the strange green leafy thing from it and taking a sip, would I really say, 'Wow, call Heinz immediately, we're onto a winner here'? I doubt it. A more likely reaction would be 'Hmm, don't give up the day job.' Yet millions of people† drink this peculiar substance every week – we like it because it's popular in Japan. Scarcity and ubiquity can both matter, depending on the context.

While in physics the opposite of a good idea is generally a bad idea, in psychology the opposite of a good idea can be a very good idea indeed: both opposites often work. I was once asked to improve a two-page letter selling an insurance product. Paragraphs had gradually been added, each of which seemed to improve the response to it – the number of sales had gradually increased. How could I improve the letter? I suggested that it be rewritten so that

---

* In Western countries at any rate, Asia seems to be different in this, to some degree.

† Including me, weirdly.

it contained no more than seven or eight lines of text. My reasoning? It was an inexpensive and sensible product, being sold by a financial company with whom the customer already had a relationship. My argument was that this simple product could be explained and understood quickly. A short letter would convey that this was a no-brainer. The existing letter, which had grown to a disproportionate length, was in danger of creating confusion‡ – if this product was as simple and sensible as it really seemed, why were they selling it so hard? We tested a two-paragraph letter. Fortunately, I was right. What had emerged was that there were two ways to sell this product: with a very long letter – which was reassuring because it was long, and with a very short letter – which was reassuring because it was very short.

The two categories of retailer who have weathered the global economic instability best in recent years are those at the top end of the price spectrum and those at the bottom. Some of this is a result of widening wealth inequality, but a glance at the demography of shoppers shows that it is not quite that simple; for instance, the bargain department store TK Maxx has a customer base that perfectly matches the UK population.§ In fact, we derive pleasure from 'expensive treats' and also enjoy finding 'bargains'. By contrast, the mid-range retailer offers far less of an emotional hit; you don't get a dopamine rush from mid-market purchases.

I was reminded of this idea recently when my wife and I were buying bed linen. After wandering around a department store for half an hour, I explained that there were only two sums of money I was prepared to spend in the store: 'zero' or 'a lot'. Zero would be good, as we could keep our existing linen and spend the money on other things. A lot of money was also acceptable, as I could then

‡ The technical term is 'cognitive dissonance'.

§ Even the super-rich love a bargain. In fact supermarket own-brand products tend to be bought more by wealthier people than by poorer people.

become excited by thread counts, tog ratings and exotic goose down. By contrast, spending something in between would have given me neither of these two emotional rewards.

The success of the brilliant engineer-alchemist James Dyson in selling vacuum cleaners seems to arise from a similar mental disparity. Vacuum cleaners used to be a grudge buy that was only necessary when your old one had broken. Dyson added a degree of excitement to the transaction. Before he invented them, there was no public clamour for 'really expensive vacuum cleaners that look really cool', any more than people before Starbucks were begging cafés to sell really expensive coffee.

CONTEXT IS EVERYTHING

People are highly contradictory. The situation or place in which we find ourselves may completely change our perception and judgement. As a good illustration of this, one reliable way to lose money is to go on holiday to some exotic locale, fall in love with the local speciality alcoholic drink and decide to import it to your home country. I once heard of someone who fell in love with a banana liqueur in the Caribbean and bought the right to sell it in the UK. On returning home with his suitcase half full of the stuff, he opened a bottle in his kitchen, hoping to impress his friends with his astute decision. Everyone, including him, found the drink practically vomit-inducing; it had only tasted good when he was in the Caribbean.*

Our very perception of the world is affected by context, which is why the rational attempt to contrive universal, context-free laws

* Pernod, of course, only tastes really good in France. And Guinness tastes better in Ireland. But that's not because Guinness is better in Ireland, but because Ireland is a better backdrop for drinking Guinness. Apparently rosé wine tastes much better if you are by the sea.

for human behaviour may be largely doomed.† Even our politics seems to be context-dependent. For instance, ostensibly right-wing people will engage – at a local level – in behaviour that is effectively socialist. A Pall Mall club in London is typically full of rich, right-wing people, yet everyone pays equal membership fees, even though they use the club in wildly different ways. Goldman Sachs, as the author and philosopher Nassim Nicholas Taleb points out, is surprisingly socialistic internally: people distribute their gains among a partnership. However, no one there proposes a profit share with JP Morgan; in one context people are happy to share and redistribute wealth, but in another, they definitely aren't.

Why is this? In his book *Skin in the Game* (2018), Taleb includes what might be the most interesting quotation on an individual's politics I have ever read. Someone‡ explains how, depending on context, he has entirely different political preferences: 'At the federal level I am a Libertarian. At the state level, I am a Republican. At the town level, I am a Democrat. In my family I am a socialist. And with my dog I am a Marxist – from each according to his abilities, to each according to his needs.'

In solving political disputes 'rationally' we are assuming that people interact with all other people in the same way, independent of context, but we don't. Economic exchanges are heavily affected by context and attempts to shoehorn human behaviour into a single, one-size-fits-all straitjacket are flawed from the outset – they are driven by our dangerous love of certainty: However, this can

† In understanding the folly of seeking universal laws for human behaviour, I have been greatly enlightened by the anthropologist Oliver Scott Curry and the recent book *Skin in the Game* by Nassim Nicholas Taleb. The attempt by philosophers to impose context-free moral obligations on people seems to fall foul of our evolved nature.

‡ One of the brothers Geoff and Vince Graham.

only come from theory, which by its very universal nature doesn't take context into account.

Adam Smith, the father of economics – but also, in a way, the father of behavioural economics[§] – clearly spotted this fallacy over two centuries ago. He warned against the 'man of system', who:

'is apt to be very wise in his own conceit; and is often so enamoured with the supposed beauty of his own ideal plan of government, that he cannot suffer the smallest deviation from any part of it. He goes on to establish it completely and in all its parts, without any regard either to the great interests, or to the strong prejudices which may oppose it ... He seems to imagine that he can arrange the different members of a great society with as much ease as the hand arranges the different pieces upon a chess-board. He does not consider that the pieces upon the chess-board have no other principle of motion besides that which the hand impresses upon them; but that, in the great chess-board of human society, every single piece has a principle of motion of its own, altogether different from that which the legislature might chuse [*sic*] to impress upon it. If those two principles coincide and act in the same direction, the game of human society will go on easily and harmoniously, and is very likely to be happy and successful. If they are opposite or different, the game will go on miserably, and the society must be at all times in the highest degree of disorder.'

The irony is that the 'man of system' in the early twenty-first century is all too likely to be an economist, but what we need more of today is men and women who are *not* wedded to an overbearing system of thought. This book is an attempt not only to create them, but also to give them permission to act and speak more freely. I hope it will free you slightly from the modern rationalist straitjacket, and help you understand that many problems might be solved if

§ Before *The Wealth of Nations*, Smith wrote a book called *The Theory of Moral Sentiments* (1759). Commonly described as a work of moral philosophy, it is also a fabulous primer on behavioural science and consumer psychology. See page 201.

we abandoned the rationalist obsession with universal, context-free laws. Once free of this constraint, you might have the freedom to generate magical ideas, some of which may be silly but of which others will be invaluable.

Unfortunately, many of your friends and colleagues, and most of all your finance director or your bank manager, won't like any of these new non-sensical ideas, even the valuable ones. That's not because they are expensive – most of them are very cheap indeed. No, he¶ will hate them because they don't sit comfortably with his narrow, reductive worldview. But that's the whole point – his narrow economic worldview has dominated decision-making for far too long.

With just a few lessons from behavioural economics and a bit of evolutionary psychology, you'll soon see where this logical world-view comes dangerously unstuck. Meanwhile your finance director, lovely guy though he may be, hates experiments involving alchemy because alchemy works erratically; he prefers small *certain* gains to those which on average will be higher but where the payoff is hard to calculate in advance.**

However, this natural human love of certainty may also prevent businesses from making more valuable discoveries. After all, no big business idea makes sense at first. I mean, just imagine proposing the following ideas to a group of sceptical investors:

1. 'What people want is a really cool vacuum cleaner.' (Dyson)
2. '... and the best part of all this is that people will write the entire thing for free!' (Wikipedia)
3. '... and so I confidently predict that the great enduring fashion of the next century will be a coarse, uncomfortable fabric which fades unpleasantly and which takes ages to dry. To date, it has been largely popular with indigent labourers.' (Jeans)

¶ And it's usually a he, isn't it?

** Why do you think Management Consultancies are so successful?

4. '... and people will be forced to choose between three or four items.' (McDonald's)
5. 'And, best of all, the drink has a taste which consumers say they hate.' (Red Bull)
6. '... and just watch as perfectly sane people pay $5 for a drink they can make at home for a few pence.' (Starbucks)[††]

No sane person would have invested a penny in these schemes. The problem that bedevils organisations once they reach a certain size[‡‡] is that narrow, conventional logic is the natural mode of thinking for the risk-averse bureaucrat or executive. There is a simple reason for this: you can never be fired for being logical. If your reasoning is sound and unimaginative, even if you fail, it is unlikely you will attract much blame. It is much easier to be fired for being illogical than it is for being unimaginative.

The fatal issue is that logic always gets you to exactly the same place as your competitors. At Ogilvy, I founded a division that employs psychology graduates to look at behavioural change problems through a new lens. Our mantra is 'Test counterintuitive things, because no one else ever does.' Why is this necessary? In short, the world runs on two operating systems. The much smaller of them runs on conventional logic. If you are building a bridge or building a road, there is a definition of success that is independent of perception. Will it safely take the weight of X vehicles weighing Y kg and travelling at Z mph? Success can be defined entirely in terms of objective scientific units, with no allowance for human subjectivity.[§§]

†† And please don't get me started on bottled water!

‡‡ Typically the size at which they start employing market research companies.

§§ Although a Swiss genius called Robert Maillart *did* build bridges based on subjective judgement. Maillart's bridges would all rank among the 100 most beautiful bridges in the world. Google them and judge for yourself. Maillart was not really an engineer – he was an artist in concrete.

This may be true when you are building a road, but it is not true when you are painting the lines on it. Here, you have to consider the more complex component of how people respond to informational cues in their environment. For instance, if you want vehicles to slow down, painting parallel lines across the road in the approach to a junction at increasingly smaller intervals will help, since the narrowing gaps between the lines will create the sensation that the car is slowing less than it really is.

Americans aren't terribly good at designing roundabouts, or 'traffic circles' as they call them, simply because they don't have much practice.¶¶ In one instance, a British team was able to reduce the incidence of accidents on a traffic circle in Florida by 95 per cent by placing the painted lines differently. In one Dutch town traffic experts improved traffic safety by removing road markings altogether.***

So there are logical problems, such as building a bridge. And there are psycho-logical ones: whether to paint the lines on the road or not. The rules for solving both are different; just as I make a distinction between nonsense and non-sense, I also use a hyphen to distinguish between logical and psycho-logical thinking. The logical and the psycho-logical approaches run on different operating systems and require different software, and we need to understand both. Psycho-logic isn't wrong, but it cares about

¶¶ We have loads of practice, since we have more roundabouts than any other country bar France. Indeed we invented the roundabout, but since this post-dated the War of Independence by 150 years or so, we failed to get the United States to show much interest. Other British possessions were more fortunate: the Swahili for 'roundabout' is 'keepi lefti' – since Kenyan roundabouts were usually marked by a sign exhorting drivers to 'Keep Left'.

*** The opposite of a good idea can also be a good idea, remember!

The initial roundabout design in Clearwater was not helped by placing a vast decorative fountain in the middle of it. A later redesign reduced the accident rate significantly.

different things and works in a different way to logic. Because logic is self-explanatory, our preference is to use it in all social and institutional settings, even where it has no place. The result is that we end up using inappropriate software for the operating system, neglecting the psycho-logical approach.

THE FOUR S-ES

There are five main reasons why we have evolved to behave in seemingly illogical ways, and they conveniently all begin with the letter S.* They are: Signalling, Subconscious hacking, Satisficing and Psychophysics. Without an understanding of these concepts, rational people will be condemned to spend their lives baffled and confounded by the behaviour of others; with a grasp of these principles, some of the oddities of human behaviour will start to fall into place.

---

* Except for the one that begins with a P.

WHY WE SHOULD IGNORE OUR GPS

The radio navigation system known as GPS is a masterpiece of logic, but it is psycho-logically dumb: what you want and what your GPS device thinks you want don't always coincide. GPS defines your task in mathematical, logical terms – getting to your destination as quickly as possible. Granted, distance may be a secondary variable – drivers may be annoyed (and financially worse off) if their GPS device saves 30 seconds on a journey by directing them to take a route via a faster road which is 20 miles longer, so there is a formulation to prevent this. But, other than projected average speed and distance, no other variables are considered.

GPS navigation is certainly a miraculous device and a triumph of logical thinking. A network of US military satellites more than 10,000 miles above the surface of the Earth, each broadcasting a signal with little more power than a 100-watt light bulb, enables a device in your car or phone to triangulate your location to within seven yards or so,* which means your phone or GPS can, after factoring

* The system is so finely tuned that the clocks on board these satellites must be calibrated to run 38 microseconds a day slower than Earth time, to correct for the effects of general and special relativity.

in prior and real-time traffic information, calculate the quickest route to any destination with an astonishing degree of precision.

Despite this, I still ignore the advice from my GPS quite a lot, especially if the journey is one of which I have prior experience, or if my psycho-logical priorities differ from its logical ones. This is because GPS is both incredibly clever but at the same time dogmatic and presumptuous.† It will confidently instruct you to take a particular route, based on a perfect understanding of a very narrow set of data points and a simplistic model of your motivation. It exhibits no sensitivity to context or to the varying priorities you may have. GPS devices know everything about what they know and nothing about anything else.

Furthermore, all navigation applications assume you are trying to reach your destination as quickly as possible, but I am not a piece of freight – if I am on holiday, I may wish to take a longer, more scenic route. If I am commuting home, I may prefer a slower route that avoids traffic jams. (Humans, unlike GPS devices, would rather keep moving slowly than get stuck in stop-start traffic.) GPS devices also have no notion of trade-offs, in particular relating to optimising 'average, expected journey time' and minimising 'variance' – the difference between the best and the worst journey time for a given route.

For instance, whenever I drive to the airport, I often ignore my GPS. This is because what I need when I'm catching a flight is not the fastest average journey, but the one with the lowest variance in journey time – the one with the 'least-bad worst-case scenario'. The satnav always recommends that I travel there by motorway, whereas I mostly use the back roads. My journey via A-roads usually takes 15 minutes longer than going via the M25, but I am happy to accept this, because a 15-minute delay will still see me arrive in plenty of time; it is preferable to the small-but-significant risk of

† In a less politically correct age, we might have described it as 'a bit German'.

spending an hour and a half stationary on a gridlocked M25, which would cause me to miss my flight.‡

The GPS knows only what it knows, and is blind to solutions outside its frame of reference. It is completely unaware of the existence of public transport, and so will suggest that I drive into central London at eight o'clock in the morning, a journey only a lunatic would undertake. By contrast, my Transport for London app is completely unaware of the invention of the automobile. And on Google Maps, once I click the 'public transport' button, it assumes I own no car (a very Californian assumption) and will confidently recommend that I travel to my nearby train station – an easy drive of no more than 15 minutes – by an elaborate combination of bus routes that will take an hour and a quarter.

To understand this book you have to realise that there is a duality in the human brain that is rather similar to the relationship between the logic of the GPS and the wider wisdom of the driver, between logic and psycho-logic. There is the unambiguously 'right' answer, where certainty is achieved by limiting the number of data points considered. The downside of this is that, in the wrong context, it can be hopelessly wrong. Then there is the pretty good judgement of psycho-logic, which considers a far wider range of factors to arrive at a not-perfect-but-rarely-stupid conclusion.

‡ Motorways are high on optimality but low on optionality: if you are stuck on a back road you can turn off and try a different route, whereas on a fast road you are trapped. A GPS understands none of this, but a human instinctively does. Nassim Nicholas Taleb's book *Antifragile* (2012) is a masterclass in understanding these second-order considerations. For instance, having someone drop 120 pebbles on your head, one every minute for two hours, would be irritating; having someone drop a rock on your head once is fatal: 1 x 120 does not always equal 120 x 1. More about this later.

The credibility we should attach to these two modes of thinking varies according to context. Sometimes it is best to obey your GPS slavishly, but at other times you should ignore it completely and use the wider parameters of judgement. Once again, the reason we don't always obey our GPS is not because we are wrong: it is because there are important factors in our journey-planning that the GPS is completely ignorant of. A lot of supposed 'irrationality' can be explained by this.

The reason we don't always behave in a way which corresponds with conventional ideas of rationality is not because we are silly: it is because we know more than we know we know. I did not decide to travel to the airport by back roads because I had calculated the level of variance in journey time – I did it instinctively, and was only aware of my unconscious reasoning in retrospect. 'The heart has reasons of which reason knows nothing,' as Pascal put it.§

However, in some cases, our conscious and unconscious decision-making coincide: on the way home from the airport, when I have no time pressure, I will generally follow my GPS. At other times we make reason take a back seat. If I were travelling down the Loire Valley, I would probably turn off my GPS and consult a decent guidebook instead; my GPS, if possessed of consciousness, would think I was a complete idiot, crawling down slow roads and crossing narrow bridges past various chateaux, when there was an autoroute just a few miles away.

In fact, my GPS goes bonkers even if I pull off the main road to fill up with fuel – 'Make a U-turn ... Make a U-turn ... MAKE A U-TURN!' It has a very narrow conception of what I am trying to do. But driving down the Loire, I would attach a low priority to the speed at which I reach my destination – a GPS simply cannot understand a motivation like this. It understands time, speed and distance, but it doesn't really have any metrics for architectural magnificence.

---

§ *'Le cœur a ses raisons que la raison ne connaît point.'* Basically seventeenth-century French for 'Sometimes it pays to ignore your GPS.'

Just as your GPS has not yet been configured to understand a wider set of human motivations, our conscious brain has not evolved to be aware of many of the instinctive factors that drive our actions. A fascinating theory, first proposed by the evolutionary biologist Robert Trivers and later supported by the evolutionary psychologist Robert Kurzban, explains that we do not have full access to the reasons behind our decision-making because, in evolutionary terms, we are better off not knowing; we have evolved to deceive ourselves, in order that we are better at deceiving others. Just as there are words that are best left unspoken, so there are feelings that are best left unthought.¶ The theory is that if all our unconscious motivations were to impinge on our consciousness, subtle cues in our behaviour might reveal our true motivation, which would limit our social and reproductive prospects.

Robert Trivers gives an extraordinary example of a case where an animal having conscious access to its own actions may be damaging to its evolutionary fitness. When a hare is being chased, it zigzags in a random pattern in an attempt to shake off the pursuer. This technique will be more reliable if it is genuinely random and not conscious, as it is better for the hare to have no foreknowledge of where it is going to jump next: if it knew where it was going to jump next, its posture might reveal clues to its pursuer. Over time, dogs would learn to anticipate these cues – with fatal consequences. Those hares with more self-awareness would tend to die out, so most modern hares are probably descended from those that had less self-knowledge. In the same way, humans may be descended from ancestors who were better at the concealment of their true motives. It is not enough to conceal them from others – to be really convincing, you also have to conceal them from yourself.

I think Robert Trivers is right in his theory of self-deception; if he were not, our job as advertisers would be much easier than it is. We

¶ Such as, 'I bought you these flowers in the desperate hope that you would sleep with me.' Or, 'I am very eager to study History of Art at your venerable Oxford college so that I can impress the recruitment panel at JP Morgan.'

could just ask people why they did things or whether they would buy them, and they would reply honestly: 'No I wouldn't normally pay $4.65 for a coffee, but if you put a fancy green logo on a paper cup so I could display it to everybody as I walk into the office then I might just be interested ...' In reality, no one will ever tell you that.

The late David Ogilvy, one of the greats of the American advertising industry and the founder of the company I work for, apparently once said, 'The trouble with market research is that people don't think what they feel, they don't say what they think, and they don't do what they say.'** Trivers and Kurzban explained the evolutionary science behind that conundrum: we simply don't have access to our genuine motivations, because it is not in our interest to know. Here's Ogilvy's contemporary, Bill Bernbach:

'Human nature hasn't changed for a million years. It won't even change in the next million years. Only the superficial things have changed. It is fashionable to talk about the changing man. A communicator must be concerned with the unchanging man – what compulsions drive him, what instincts dominate his every action, even though his language too often camouflages what really motivates him.'

Years ago, in an interview on online book-buying for a client, a young man told me something surprisingly honest. 'Look, to be frank, I don't like reading novels all that much, but I find if you have read a few Ian McEwan [novels] you can pull a much better class of girl.' Such candour about our deeper motives is rare.††

Human self-deception makes our job difficult for another reason: no one wants to believe in its existence, and it is something which

---

** I can find no evidence that Ogilvy did say this – he started his career in market research, and was a great promoter of it. But I nonetheless think he would have reluctantly agreed with it.

†† It is not clear that his self-awareness was much help to his reproductive prospects: I seem to remember he was single. Perhaps his motivation was too obvious to the girls he met?

people seem only to accept at a shallow, theoretical level.[‡‡] People are much more comfortable attributing the success of a business to superior technology or better supply-chain management than to an unconscious, unspoken human desire.

Perhaps that's because we patently need a level of self-delusion to function as a social species.[§§] Imagine a world where we had no capacity for deception, and where people on dates directly asked prospective partners about their earning power and career prospects, without even pretending to be interested in their personalities. Where would we be then?[¶¶]

Evolution does not care about objectivity – it only cares about fitness.

If it helps us to perceive the world in a distorted fashion, then evolution will limit our objectivity. The standard, naïve view, as Trivers observes, is to assume that evolution has given us senses which deliver an accurate view of the world. However, evolution cares nothing for accuracy and objectivity: it cares about fitness. I may know rationally a snake is harmless, but instinctively I'm still unnerved by the slithery bastards.

It isn't easy to get people to accept the idea of hidden motivations. After all, cat lovers might realise that their pet tends to get more affectionate when it is hungry, but good luck getting them to believe that their beloved moggy is only faking affection to get food. Nevertheless, we would all benefit if we learn to accept the fact that our unconscious motivations and feelings may have remarkably little to do with the reasons we attribute to them.

---

†† Even people working in advertising, to be honest.

§§ Just as, in evolutionary terms, it may pay us to be over-optimistic, rather than objective about our prospects. In psychology, intriguingly, the only people found not to suffer from overconfidence-bias are people with severe depression.

¶¶ New York, perhaps.

Remember the airline and the cucumber sandwiches? Just as we infer a great deal about an air carrier from their on-board catering, while neglecting to care about the $150m aircraft or the make of the engines, we are just as likely to be unhappy with a hospital because the reception area is neglected, the magazines are out of date and the nurse didn't spare us much time. In truth, the UK's National Health Service might benefit from 'wasting' a bit more money on signalling, while the US healthcare sector could probably benefit from spending a lot less. It is fine to provide up-to-date magazines in reception to show that you care, but when the urge to show commitment to patients involves performing unnecessary tests and invasive surgery, it probably needs to be reined back.

Research will never tell you this; if surveyed, we would insist that the objective health measures are all we care about, and we would believe what we are saying. But the truth is that ancillary details have a far greater effect on our emotional response, and hence our behaviour, than measured outcomes. Consider these contrasting statements: 'She died yesterday, but I must say the hospital was wonderful.' Or 'No, Dad's fine. No thanks to that bloody hospital, mind. He was kept waiting four days for his operation.' Objectively, the UK NHS provides very good medical outcomes for the money spent on it; the sad result is that we don't like it any more than we would enjoy a flight on a brand new airliner whose sandwiches were starting to curl.

For a business to be truly customer-focused, it needs to ignore what people say. Instead it needs to concentrate on what people feel.

Let me give you an example of how ignoring what people say can be creatively liberating. Like the question of how we assess hospitals and medical care, it concerns the question of emotional misattribution; the fact is that, while we know how we feel, we cannot accurately explain *why*. Nature cares a great deal about feelings, and feelings largely drive what we do, but they do not come with explanations attached – because we are often better off not knowing them.

What we think about how we feel may have little to do with our real reasons for feeling it, so it pays often to ask naïve questions to which the answers seem completely self-evident. 'Why do people go to restaurants?', say. 'Because they are hungry,' comes the answer. But if you think about it a little, someone merely hungry could satisfy their urge to eat far more economically elsewhere. Restaurants are only peripherally about food: their real value lies in social connection, and status.***

It's interesting that, once we leave childhood, we stop asking these apparently childish questions. Try this exercise, which starts with a childish question, but one that might not have been asked before: Why don't people like being made to stand on overcrowded trains?††† I once asked this question in a meeting with a rail company. Everyone looked nonplussed; I mean, it's obvious that standing has to be worse than sitting, right? Maybe so. But why? And if standing is always worse than sitting, why do people standing on trains regularly continue to stand after seats become available? There could be a whole variety of reasons but, fascinatingly, passengers themselves do not really know, even if they are able to supply plausible post-rationalisations. But asking this question more broadly might lead to interesting new railway carriage designs that nobody has yet thought of, or it might be solved by differential pricing. We don't know yet.

So let's ask again – why might people hate standing on trains? Is it about feeling cheated? After all, you've paid for a seat on the train, and the rail company has taken your money and not given you a seat. Is that it? In which case, might you try offering standing-only carriages for shorter rail and tube journeys? People using them

*** Tellingly, perhaps, most of their money is made selling not food, but alcohol.

††† I am talking about the London Underground and commuter trains here. The question 'Why do people mind standing for a four-hour journey?' *would* be childish!

could be refunded part of their fare, or rewarded with points towards free journeys. Would they feel happy then? We could find out.

Or perhaps it's because it is tiring; it's not just about having to stand, it's also about having to keep your balance. Or that, once you have to hold on to a pole to stay upright, you can no longer use a mobile phone, read a book or newspaper or drink a coffee, so the journey becomes boring. If these are the reasons, then a series of bum-rests might help.‡‡‡ Perhaps it's because they have nowhere to put their bags or they are paranoid about people stealing from their backpack.§§§ Maybe though, it's more a question of status; the people who have a seat have a view, control of their personal space and space for their bags – while the people who stand get nothing. There is no story they can tell themselves about their predicament that puts it in a better light. But this raises an interesting question: what if there were some benefits to standing? In other words, is there a role for alchemy?

Imagine if commuter rail carriages were designed with the seats down the middle, with places for passengers to stand down each side, next to the windows. People sitting might have cup-holders but nothing else; people standing would have a view out of the window, a cushion to rest against and a shelf for a bag or a laptop, with two USB charging sockets. Now there would be some clear advantages to standing over sitting, to a point where standing could be perceived – by others and by oneself – as a choice rather than a compromise.¶¶¶

‡‡‡ Such padded rests do exist at the end of London Underground carriages, and people sitting on them never seem unhappy.

§§§ Interestingly a British company has just launched a backpack with all the zips facing inwards towards your back, precisely to solve this fear.

¶¶¶ 'A choice, not a compromise' was at one time Ogilvy's slogan for the Ford Fiesta. Advertising lines – 'reassuringly expensive' for Stella Artois, say, often unintentionally offer useful insights into psycho-logic.

Plans such as this only emerge when people ask a dumb question with an open mind. The commuter knows he hates standing, but he does not really know why; if you ask him, he will demand more seats, but the only way to provide them is through the huge expense of running more trains. The reason we do not ask basic questions is because, once our brain provides a logical answer, we stop looking for better ones; with a little alchemy, better answers can be found.

PART 1: ON THE USES AND ABUSES OF REASON

SOMETHING CLEARLY WENT WRONG with food in both Britain and America between the 1950s and the 1980s: it came to be considered to be more about convenience than pleasure. It seems astonishing now, but the predictions of the future I read as a child assumed that meals would be replaced by parcels of nutrients consumed in handy tablet form – it was for some reason thought that the purpose of food was to provide the necessary minerals, vitamins, protein and energy, and that the job of the food industry was to supply them in as efficient a form possible.

Some forward-thinking people had defined food's function narrowly, in order to create a rational model of what the food industry should do.* In this focus on scale and efficiency, people lost sight of what food is *for*; while it is, of course, a form of nourishment, it also serves a host of other ends. The proponents of delivering food in pill form had lost sight of the fact that it is enjoyable to eat and a necessary prop at social occasions.† Even if such pills could be

* If this seems ridiculous in retrospect, remember that Silicon Valley may frequently be doing this same thing today: destroying variety and pleasure in pursuit of a logical end that would be psychologically disastrous.

† If you attend a meeting with the UK Government, no biscuits are provided. It saves something like £50m a year. The hidden cost is that every meeting takes on a

produced, it is perfectly plausible that people who ate only such food would be utterly miserable.

In many ways it is the very inefficiency of premium foods that gives them their emotional value. The sourdough bread beloved of hipsters is insanely slow and inefficient to produce. Likewise it is absurd for the French to have so many local varieties of cheese, and yet this variety and scarcity seems to add to our pleasure. Contrast it with the US cheese industry thirty years ago – which was fabulously efficient and centred on a small number of states. In the 1990s there seemed to be only two varieties of cheese, a yellow one and an orange one, and neither was much good. Similarly, before the recent revolution in craft beer, the range and quality of American beer was dismal;[‡] however, since American brewing has become magnificently diverse and inefficient, the US has gone from being the worst country for a beer drinker to visit, to the best.[§]

Food has become remarkably inefficient, and the pill-promoting futurists of the 1960s would be astonished to see how wrong they were. People spend hours preparing it, eating it and watching television programmes about it. People cherish local ingredients, and willingly pay a premium for foods produced without chemical fertilisers. By contrast, when we made the food industry logical, we lost sight of the reasons we value food at all.

Using this as a metaphor, I would like to see the improvement we have enjoyed in food over the last three decades applied to

slightly unpleasant timbre by violating the most basic principles of hospitality. I don't even like biscuits, but it still pisses me off. Chatting in the absence of biscuits feels less like a cooperative meeting and more like being interrogated by a Serbian militia. Under any Sutherland regime, scones will be mandatory.

‡ Wisconsin, I'm afraid to say, has to carry the can for both the cheese and the beer.

§ A US craft brewery has recently opened in Germany.

other fields. It is only when we abandon a narrow logic and embrace an appreciation of psycho-logical value, that we can truly improve things. Once we are honest about the existence of unconscious motivations, we can broaden our possible solutions. It will free us to open up previously untried spaces for experimentation in resolving practical problems if we are able to discover what people really, really want,¶ rather than a) what they say they want or b) what we think they should want.

¶ Zigazig ah! (Joke comprehensible to Spice Girls fans only).

1.1 THE BROKEN BINOCULARS

For the last fifty years or so, most issues involving human behaviour or decision-making have been solved by looking through what I call 'regulation-issue binoculars'. These have two lenses – market research and economic theory – that together are supposed to provide a complete view of human motivation. There's only one problem: the binoculars are broken. Both the lenses are pretty badly cracked, and they distort our view of every issue.

The first lens is market research or, to give it a simpler name, asking people. However, the problem with it is that, if we remember David Ogilvy's words: 'The trouble with market research is that people don't think what they feel, they don't say what they think, and they don't do what they say.' People simply do not have introspective access to their motivations. The second lens is standard economic theory, which doesn't ask people what they do and doesn't even *observe* what they do. Instead it assumes a narrow and overly 'rationalistic' view of human motivation, by focusing on a theoretical, one-dimensional conception of what it *believes* humans are trying to do. Again, behavioural economics has shown that it provides an incomplete and sometimes misleading view of human behaviour – neither the business nor the policy worlds have paid sufficient attention to the failings of economics and research. Why might this be?

Generally, it is safe for anyone making business or policy decisions to act as though everything seen through these binoculars is accurate – not least because everyone else they work with – and everyone who might hire, promote or fire them – sees the world through the same binoculars.

'The economic model told me to do it' is the twenty-first-century equivalent of 'I was only following orders,' an attempt to avoid blame by denying the responsibility for one's actions. Sometimes the old binoculars work well, of course: quite often people can accurately describe their motivations, and a large part of human behaviour is perfectly consistent with economic theory. Logic and psycho-logic do overlap frequently, as you would expect.

However, we still need a new set of lenses; as I explained at the beginning of the book, stubborn problems are probably stubborn because they are logic-proof. In other cases, the old binoculars provide a view that is so distorted, a field of view so narrow, that they blind us to far simpler creative solutions. The broken binoculars assume that the way to improve travel is to make it faster, that the way to improve food is to make it cheaper and that the way to encourage environmentally friendly behaviour is to convert people into passionate environmentalists. All these ideas are sometimes true – but not always.

Any new binocular lenses provided by sciences such as behavioural economics and evolutionary psychology will not be flawless, but they can at least provide us with a wider field of view. All progress involves guesswork, but it helps to start with a wide range of guesses. The following is a simple example of how a new lens can allow you to see (and solve) problems from a more psycho-logical perspective.

One of our clients at Ogilvy Change is a large energy provider who arranges appointments with customers for engineers to repair or service their central heating boilers. The appointments are scheduled either for the morning or for the afternoon – it is difficult to be any more precise than this, since it is hard to predict how long each visit may take. Customers complain about this, their most common refrain being, 'I had to take the whole day off work.' What these customers *say* they want is a one-hour appointment window. However, if you were to take their demands literally and attempt such a level of precision, it would cost a fortune and there would be a risk of disappointment whenever circumstances prevented an engineer from delivering on the promise. The more astute of you may also have noticed that the one-hour appointment window would not necessarily solve the problem of 'having to take a day off work' – if your appointment was between 1pm and 2pm, for instance, unless you worked a short distance from home, you'd still have to take a day off work to be available.

Our first recommendation to the client was to listen to what consumers said, but to interpret it *laterally* rather than *literally*. People clearly found something about the length of the appointment window annoying, but maybe it was the degree of uncertainty involved in waiting for the engineer to show up rather than the length of the appointment window. Anyone who has waited at home for five hours for an engineer knows that it's a form of mental torture, a little like being under house arrest; you can't have a bath or pop out for a pint of milk, because you fear that the second you do, the engineer will turn up. So you spend half the day on tenterhooks, afraid that your engineer might not show up at all. How different might the experience feel if the engineer agreed to text you half an hour before showing up at your door? Suddenly you'd be free to get on with your day almost as if it were a day off, with your only obligation being to keep an eye on your phone.* This is one of the solutions we propose to test. Is it as good as offering one-hour appointments? Not quite, but it might offer 90 per cent of the emotional and perceptual improvements, at 1 per cent of the cost. The

* There is quite a lot of behavioural evidence to support our assertion. For instance, accurate, real-time

old binoculars would not have revealed this because they would have taken customer complaints literally.

My colleague Christopher Graves, who founded the Ogilvy Center for Behavioral Change in New York, calls this approach 'asking the real why'. People may be accurate commentators on their emotional state, but the causes of that emotional state (in this case, uncertainty) are often a complete mystery to them. If the experiment works, and early indications are positive, we have performed a form of alchemy, using psycho-logic to conjure up value from nowhere. Experimentation is the only reliable way of testing, so we measure the effect of engineers' texts on customer satisfaction against a control group who receive no such early warning.

Another method is to perform what is called a thought experiment. For instance, ask yourself which message on a flight departure board would distress you more:

BA 786 – Frankfurt – DELAYED

or

BA 786 – Frankfurt – DELAYED 70 minutes.

The second message is a bit of a pain – but at least you are in control of the situation. You may need to make a few apologetic telephone calls, or go to a lounge and get your laptop out, but you can get on with re-planning your day. The first message, however, is a form of mental torture. You know there is bad news, but you do not have sufficient information to respond to it. Is it a 10-minute delay or a 90-minute delay? You might also worry that 'delayed' is merely a precursor to 'cancelled'. That loss of power and

departure boards at train stations, which do not make your journey any faster, add a great deal to passenger satisfaction – it seems we would rather wait eight minutes for a train knowing it will arrive in eight minutes, than spend four minutes waiting for a train in a state of anxious uncertainty.

control can create far stronger feelings of annoyance than the loss of punctuality.†

Unfortunately we are unable to distinguish between these two emotions: you don't say, 'I am unhappy because inadequate information has left me powerless'; you say, 'I'm angry because my bloody plane's late.' In such cases, neither lens of the binoculars will present you with a solution. Airline passengers won't want me to say this, but it's true: if, as an airline, you have a choice between delaying a flight by an hour or spending £5,000 to leave on time, your decision should be influenced by the quality of passenger information you can provide. I would also say that, from a psycho-logical point of view, metrics which target the punctuality of an airline without factoring in the quality of their information may be encouraging them to optimise the wrong thing.‡ (Remember also that perhaps twenty passengers on any flight might be delighted to receive a text message telling them that their flight is delayed – namely the people running late.)§

† If you are interested in understanding more about the depressive effect of feeling a lack of control over an aversive stimulus, or the negative psychological effects of bad design, investigate the work of Don Norman (*The Design of Everyday Things*, 1988) and the 'learned helplessness' experiments of Martin Seligman and Steven Maier (1967).

‡ It also seems dumb from a psychological point of view not to measure delays as a percentage of the flight duration. A 30-minute delay on a one-hour flight is a far greater annoyance than a one-hour delay on a nine-hour flight.

§ Or the ones who want an excuse to cancel their bloody meeting in Frankfurt.

This might all sound like rather a trivial use of behavioural science. But, as you will learn later, the same techniques which can solve minor problems can also be deployed to solve much larger ones. For instance, the technique which might solve the problem of appointments for heating engineers may reduce people's reluctance to save for their pension.** One of the reasons I believe there is genuine value to the study of behavioural science is that the same patterns recur: a solution which at a relatively trivial level helps encourage people to apply for credit cards might also be used to make people less reluctant to have medical tests.

More on this later ...

** I will explain how later in this book.

1.2 I KNOW IT WORKS IN PRACTICE, BUT DOES IT WORK IN THEORY? ON JOHN HARRISON, SEMMELWEIS AND THE ELECTRONIC CIGARETTE

The approach that I am proposing will help you generate new and interesting ideas that are worthy of experimental testing, but do not expect them to be immediately popular or easy to sell. If you would like an easy life, never come up with a solution to a problem that is drawn from a field of expertise other than that from which it is assumed the solution will arise. A few years ago, my colleagues produced an extraordinary intervention to reduce crime. They hypothesised that the presence of the metal shutters that shops in crime-ridden areas covered their windows with at night may in fact *increase* the incidence of crime, since they implicitly communicated that this was a lawless area.

One of my colleagues, the brilliant Tara Austin, had seen research that suggested that 'Disney faces' – large-eyed human faces with the proportions of young children – seemed to have a calming effect. Combining the two ideas, she created an experiment where shop shutters were painted with the faces of babies and toddlers by a local graffiti artist collective.

By all measures, this seemed to reduce crime significantly; moreover, it did so at a tiny cost, and certainly by less than the cost of direct policing. Several other local authorities have since repeated the approach, though take-up is low – it is much easier to argue for larger policing budgets or for the installation of CCTV, than to approach a problem psycho-logically.

Contrast this effect with the impression given by plain steel shutters.

In a sensible world, the only thing that would matter would be solving a problem by whatever means work best, but problem-solving is a strangely status-conscious job: there are high-status approaches and low-status approaches. Even Steve Jobs encountered the disdain of the nerdier elements of the software industry – 'What does Steve do exactly? He can't even code,' an employee once snootily observed.

But compared to an eighteenth-century counterpart, Jobs had it easy. In the mid-eighteenth century, a largely self-taught clockmaker called John Harrison heard that the UK Government had pledged £20,000 – several million pounds in today's currency – as a prize for anyone who could establish longitude to within half a degree* after a journey from England to the West Indies, and was determined to find a solution. This was a life and death matter – a navigational disaster by British naval ships off the Isles of Scilly in 1707 had left several thousand sailors dead. To judge proposed solutions, the crown established a Board of Longitude, consisting of the Astronomer Royal, admirals and mathematics professors, the Speaker of the House of Commons and ten Members of Parliament.

You'll notice that there were no clockmakers on the committee – the prize was clearly offered under the assumption that the solution would be an astronomical one, featuring celestial measurement and advanced calculation. In the end, Harrison produced an astonishing series of discoveries that led to the invention of the marine chronometer, and with it a revolution in navigation. Once ships could carry an accurate timepiece at sea, they were finally able to calculate how far they had travelled from east to west without recourse to less reliable methods.†

Alongside Harrison's remarkable technological work, there is also an interesting psychological aspect to this story. Though they awarded him a great deal of money for his invention, the prize was always denied to him, even though he demonstrated that his solution worked more than once. A great part of his later life was spent petitioning the authorities and complaining that he had been cheated of his reward. Nevil Maskelyne, a supporter of the 'lunar distances' method of astronomical calculation, is often portrayed

---

* Thirty nautical miles at the equator.

† We owe it to Dava Sobel and her bestselling book *Longitude* (1995) that the name of John Harrison is now widely known.

as the villain for denying Harrison the prize – and it cannot have helped Harrison's case when Maskelyne was made Astronomer Royal and a member of the prize-giving committee. But the real story is one of professional and academic hierarchy: to an astronomer, the solution of an uneducated man whose life had been spent making clocks did not seem worthy of recognition.

I am not so sure that Maskelyne was a villain and instead see him as a 'typical intellectual'. I say this because we see the same pattern in a series of significant innovations – science seems to fall short of its ideals whenever the theoretical elegance of the solution or the intellectual credentials of the solver are valued above the practicality of an idea. If a problem is solved using a discipline other than that practised by those who believe themselves the rightful guardians of the solution, you'll face an uphill struggle no matter how much evidence you can amass.

Until 1948, the Wright brothers' Flyer was displayed not in the Smithsonian, but in the Science Museum in London. This might seem strange, but for years after the bicycle shop owners from Ohio had flown their manned heavier-than-air device on North Carolina's Outer Banks, the US Government refused to acknowledge their achievement, maintaining that a government-sponsored programme had actually been first.[‡] In 1847, when Ignaz Semmelweis decisively proved that hand-washing by doctors would cut the incidence of puerperal fever, a condition that could be fatal during childbirth, he was spurned. All too often, what matters is not whether an idea is true or effective, but whether it fits with the preconceptions of a dominant cabal.[§]

‡ They had hence proved not only that a heavier-than-air machine could fly, but also that snobbery is not an exclusively British vice.

§ Semmelweis was even more cruelly treated than Harrison: he died in a lunatic asylum, perhaps having been beaten by the guards, insisting to his last breath that his theory was right. Which it was.

I had always innocently assumed that after Edward Jenner discovered a vaccination against smallpox he would have presented his findings before sitting back to enjoy the acclaim. The truth was nothing of the kind; he spent the rest of his life defending his idea against a large number of people who had profited from an earlier practice called variolation, and were reluctant to admit that anything else was better. And if you think this problem is confined to history, consider the reaction to the invention of the electronic cigarette.

The scientific establishment has been right to be sceptical about e-cigarettes – we still do not know for sure what the long-term consequences of this technology might be. But the invention of a delivery device for nicotine that recreates much of the feeling of smoking without the carcinogens which accompany burning tobacco is clearly a significant idea, and something that should be given open-minded consideration. However, from the first moment this technology appeared, the opposition was out in force. Many countries banned the devices immediately, and the World Health Organization and anti-smoking groups worldwide clamoured for their use to be banned wherever smoking was banned. Weirder still, they were also banned in many Middle Eastern countries which have almost no prohibitions on smoking. The question being asked seemed to be, 'Yes I know it works in practice, but does it work in theory?' Just as, under Maskelyne, the dominant model was astronomical rather than horological, in smoking cessation the dominant model was one of shame¶ rather than of accommodation.

If you have spent the last 20 years as a public health advisor promoting policies designed to create shame, alongside colleagues who all believe the same thing, the last thing you want to hear is 'Don't worry about that, because a bloke in China has come up with a gadget which means that the problem to which you have dedicated your life and from which your social status derives is no longer a problem any more.' Even worse, the inventor was a

¶ Or 'denormalisation'.

businessman rather than a health professional. As with Maskelyne's rejection of Harrison's marine chronometer, there were vested interests at stake. Some of the funding for anti-vaping campaigns came from large pharmaceutical companies, which saw the devices as threatening their investment in less potent quitting treatments such as patches or gum.[**] Thankfully, in some countries, a kind of common sense prevailed,[††] though the majority of public health professionals nevertheless were highly reluctant to endorse the use of electronic cigarettes – they may well have been unconsciously affected by the thought that a successful substitute for cigarettes would make their skills redundant. Maskelyne might have felt the same way: 'We don't need your spectacular astronomical knowledge any more, mate, because this clockmaker has just cracked the problem.'

The same problem is widespread in medicine. Surgeons felt challenged by keyhole surgery and other new, less invasive procedures that can be carried out with the support of radiographers, because they used skills different from those that they had spent a lifetime perfecting. Similarly, you can imagine how London black cab drivers feel about Uber. As the novelist Upton Sinclair once remarked, 'It is difficult to get a man to understand something, when his salary depends on his not understanding it.'

One of the common arguments against vaping was that it renormalised smoking, because it looks a bit like smoking. I find

---

** This is known as a 'Baptists and Bootleggers' coalition, where moralists and money-makers join forces to resist some proposed relaxation of a law. Bootleggers, for obvious financial reasons, were heavily opposed to the removal of Prohibition in the United States.

†† In Britain, credit is owed to Public Health England and ASH, an anti-smoking campaign group; in the US, a former Surgeon-General was a great advocate of the invention. I even played a small part myself, by persuading the UK Government's Behavioural Insights Team to resist a knee-jerk urge to ban them.

that quite hard to believe, frankly. Whatever you think about smoking, it does look fairly cool – a remake of *Casablanca* with the cigarettes replaced with vaporisers would be somewhat less romantic. The other argument, which I found even more implausible, was that vaping might act as a gateway drug to more serious substances. Most heroin addicts may have started with cannabis, but then, most cannabis addicts probably started with tea and coffee.

A few years ago I was puffing on an e-cigarette outside an office, when a security guard came out. 'You can't smoke here,' he shouted. 'I'm not, actually,' I replied. He went to consult his superior, reappearing a few minutes later. 'You can't use e-cigarettes here, either.' 'Why not?' 'Because you are projecting the image of smoking.' 'What, you mean insouciance?'

'Go away.' I did.

That phrase, 'projecting the image of smoking' – along with 'renormalisation' and 'gateway effect' – appears frequently in arguments for restricting vaping in public places. And, while new evidence may yet emerge to support restrictions, these reasons aren't convincing; like the security guard's response, they look to me like a desperate attempt to reverse-engineer a logical argument to suit an emotional predisposition.

As the psychologist Jonathan Haidt has shown, most moralising works in this way. We react instinctively, before hastily casting about for rationalisations. For instance, most Britons feel it is repulsive to eat dogs or even horses. If you ask them why, they will contrive a series of arguments to defend what is really a socially constructed belief, just as people with a distaste for vaping seize on the argument that non-smokers will take up e-cigarettes and then migrate to real cigarettes. The gateway effect would make sense, were the evidence for it not somewhere between negligible and non-existent. The traffic seems to flow in the opposite direction – from smoking to vaping to (in many cases) quitting altogether. According to ASH, only 0.1 per cent of e-smokers have never smoked tobacco. Only 5 per cent of children use e-cigarettes more than once a week, almost all of whom are current or ex-smokers. If it is a gateway drug, it mostly seems to be a gateway *out*.‡‡

One possible explanation for this is that smoking is not so much an addiction as a habit: that after a few years of smoking, it is the associations, actions and mannerisms we crave more than the drug itself. Hence, if you have not been addicted to smoking cigarettes, e-cigs simply don't hit the spot, just as those of us who have never been heroin addicts tend not to be all that keen on needles.

A few years ago, a High Court judge was driving home from his golf club after five or six double gin and tonics when he was pulled over by the police and breathalysed. When the machine barely registered an amber light, the police let him go – at which point, he drove back to the club and demanded that the head barman be fired for watering down the drinks. Dodgy barmen have known for years that, after one proper G&T, you can sell people tonic water in a glass lightly rinsed in gin and not only will they not notice the difference, but regular drinkers may still manifest all the effects of drunkenness – slurred speech and poor coordination – even though they have consumed almost no alcohol.§§ A similar placebo effect may mean that ersatz smoking only works for ex-smokers. If that is the case, the good news for the vaping industry is that one common objection to it can be rejected, though the bad news is that e-cigarette sales may shrink as they run out of former smokers to convert.

‡‡ Frankly, I was surprised by how low the figures were. I would have expected at least 5 per cent of non-smokers to give e-cigs a try. What's going on?

§§ However, this works only if you have spent a lot of your life drinking real G&Ts. Among heavy drinkers, the brain doesn't wait for the booze to kick in – it shortcuts straight to the expected level of drunkenness.

1.3 PSYCHOLOGICAL MOONSHOTS

Alphabet, the parent company of Google, runs a division that is now simply called 'X'. It was founded as Google X, with the aim of developing what the company calls 'moonshots'. A moonshot is an incredibly ambitious innovation; instead of pursuing change by increments, it aims to change something by a factor of ten. For instance, X funds research into driverless cars, with the explicit aim of reducing road-accident fatalities by at least 90 per cent. The argument for X is that the major advances in human civilisation have come from things that, rather than resulting in modest improvement, were game-changers – steam power versus horse power, train versus canal, electricity versus gaslight.

I hope X is successful but think that their engineers will find it difficult. We are now, in many cases, competing with the laws of physics. The scramjet or the hyperloop* might be potential moonshots, but making land- or air-travel-speeds so much faster is a

* Respectively a kind of supercharged jet engine and a form of high-speed land travel through tunnels from which all air, and hence air-resistance, has been removed.

really hard problem – and comes with unforeseen dangers.[†] By contrast, I think 'psychological moonshots' are comparatively easy. Making a train journey 20 per cent faster might cost hundreds of millions, but making it 20 per cent more enjoyable may cost almost nothing.

It seems likely that the biggest progress in the next 50 years may come not from improvements in technology but in psychology and design thinking. Put simply, it's easy to achieve massive improvements in *perception* at a fraction of the cost of equivalent improvements in *reality*. Logic tends to rule out magical improvements of this kind, but psycho-logic doesn't. We are wrong about psychology to a far grater degree than we are about physics, so there is more scope for improvement. Also, we have a culture that prizes measuring things over understanding people, and hence is disproportionately weak at both seeking and recognising psychological answers.

Let me give a simple example. The Uber map is a psychological moonshot, because it does not reduce the waiting time for a taxi but simply makes waiting 90 per cent less frustrating. This innovation came from the founder's flash of insight (while watching a James Bond film, no less[‡]) that, regardless of what we say, we are much bothered by the uncertainty of waiting than by the duration of a wait. The invention of the map was perhaps equivalent to multiplying the number of cabs on the road by a factor of ten – not because waiting times got any shorter, but because they felt ten times less irritating.

† Before we get too excited by the economic prospects of the driverless car, it is worth considering what this technology might offer to terrorists, for instance. A driverless car is effectively a cruise missile on wheels.

‡ *Goldfinger*. In the film, Bond tracks Auric Goldfinger's Rolls-Royce to his Alpine lair using an animated map installed in his Aston Martin.

And yet we spend very little money and time looking for psychological solutions, partly because, in attempting to understand why people do things, we have a tendency to default to the rational explanation whenever there is one. As we saw with Maskelyne's response to Harrison's nautical innovation, the people at the top of organisations are largely rational decision makers who are naturally disparaging of psychological solutions. But it also comes from our urge to depict our behaviour in as high-minded a way as we can manage, hiding our unconscious motivations beneath a rational facade.

We may grudgingly accept that people may have unconscious emotional motivations for preferring one brand of beer over another, but that's because we don't see beer as an essential product. Most people would acknowledge that relatively trivial considerations, such as an ad campaign or the design of a label, may have an influence on what we drink in a bar, but if you suggest to people that similar unconscious motivations could be decisive in our use of healthcare or how we choose to save for retirement, people are scandalised.

I am willing to bet that there are ten times as many people on the planet who are currently being paid to debate why people prefer Coke or Pepsi than there are being paid to ask questions like 'Why do people request a doctor's appointment?', 'Why do people go to university?' or 'Why do people retire?' The answers to these last three questions are believed to be rational and self-evident, but they are not.

1.4 IN SEARCH OF THE 'REAL WHY?' UNCOVERING OUR UNCONSCIOUS MOTIVATIONS

As I have already said, if you want to annoy your more rational colleagues, begin a meeting by asking a childish question to which the answer seems self-evident – the fact that sensible people never ask questions of this kind is exactly why you need to ask them. Remember the example I gave about asking why people hate standing on trains? When I asked that question, it seemed likely that no adult on the planet had asked that question for the last ten years – it sounded like such a stupid thing to ask.

Perhaps advertising agencies are largely valuable simply because they create a culture in which it is acceptable to ask daft questions and make foolish suggestions. My friend and mentor Jeremy Bullmore recalls a heated debate in the 1960s at the ad agency J. Walter Thompson about the reasons why people bought electric drills. 'Well obviously you need to make a hole in something, to put up some shelves or something, and so you go out and buy a drill to perform the job,' someone said, sensibly. Llewelyn Thomas, the copywriter son of the poet Dylan, was having none of this. 'I don't think it works like that at all. You see an electric drill in a shop and decide you want it. Then you take it home and wander around your house looking for excuses to drill holes in things.' This discussion perfectly captures the divide between those who believe in rational explanation and those who believe in unconscious motivation; between logic and psycho-logic.*

You will never uncover unconscious motivations unless you create an atmosphere in which people can ask apparently fatuous questions without fear of shame. 'Why do people hate waiting for an engineer's appointment?' 'Why do people not like it when their flight is delayed?' 'Why do people hate standing on trains?' All of these questions seem facile – and because of this, our rationalising brains find it dangerously easy to come up with a plausible answer. But just because there is a rational answer to something, it doesn't mean that there isn't a more interesting, irrational answer to be found in the unconscious.

'Why do people mostly buy ice cream in the summer?' seems a pretty facile question. 'Duh! To cool down on a hot day!' It certainly sounds plausible, but human behaviour tells a different story. For one thing, sunshine is a far better predictor of ice cream sales than temperature. And to confuse things further, the three countries with the highest per-capita ice cream sales in Europe? Finland, Sweden and Norway. One possible way of looking at the question might be to ask whether people need the excuse of a special occasion to justify eating ice cream. Perhaps a sunny day is Sweden is rare enough to provide the necessary licence?

Similarly, 'Why do people go to the doctor?' seems like an idiotic question, until you realise that it isn't. Is it because they are ill and want to get better? Sometimes, but there are many more motivations that lie beneath this apparently rational behaviour. Perhaps they are worried and crave reassurance? Some people just need a bit of paper to prove to their employer they were ill. A lot of people may go in search of someone to make a fuss of them. Perhaps, what people are mostly seeking is not treatment, but reassurance. The

* I'm largely with Thomas – I don't do much DIY, but I would acknowledge that my principal interest in cooking is not primarily the preparation of food: it's an excuse to buy kitchen gadgetry.

distinction matters – after all, not many people make unnecessary visits to the dentist.

If you want to solve the problem of unnecessary doctor's visits or simply to set up a system to prioritise who gets seen by the doctor first, it is vital that you factor in unconscious motivations alongside post-rationalisations. Some problems might be solved over the phone, while other visits could be postponed until it was likely the person had recovered naturally. In the event of a flu outbreak, you might even leave an answerphone message detailing the symptoms and telling younger or less vulnerable people what to do if they were suffering. Once people know an illness is widespread they are less anxious about being ill, and correspondingly less eager to see a doctor for reassurance. 'There's a lot of it about' is reassurance in itself. (What you *don't* want your doctor to say is, 'This is an extraordinary case – I've never seen anything like it in my whole professional career.')

The strange thing is that everyone is much happier pretending that the post-rationalised reason for visiting the doctor, to get better, is the only one that counts. If you want to change people's behaviour, listening to their rational explanation for their behaviour may be misleading, because it isn't 'the real why'. This means that attempting to change behaviour through rational argument may be ineffective, and even counterproductive. There are many spheres of human action in which reason plays a very small part. Understanding the unconscious obstacle to a new behaviour and then removing it, or else creating a new context for a decision, will generally work much more effectively.

Whether we use logic or psycho-logic depends on whether we want to solve the problem or to simply to be seen to be trying to solve the problem. Saving the world indirectly may not make you look like a hero; talking about the plight of polar bears makes one feel a good deal worthier than promoting the redesign of recycling bins, but the latter may be more effective. The self-regarding delusions of people in high-status professions lie behind much of this denial of unconscious motivation. Would you prefer to think of yourself as a medical scientist pushing the frontiers of human knowledge, or as a kind of modern-day fortune teller, doling out soothing remedies to worried patients? A modern doctor is both of these things, though is probably employed more for the latter than the former. Even if no one – patient or doctor – wants to believe this, it will be hard to understand and improve the provision of medical care unless we sometimes acknowledge it.

1.5 THE REAL REASON WE CLEAN OUR TEETH

There is one example of a human behaviour which has both an 'official' medical purpose and a deep psychological explanation, and which I think is helpful in showing how a logical, rational explanation for our behaviour may drown out the unconscious, evolutionary one. It starts with another childish question, 'Why do people clean their teeth?' Obviously it is to maintain dental health and to prevent cavities, fillings and extractions. What possible other answer could there be? Well, in fact, if we look at adult behaviour – when we choose, buy and use toothpaste – we see patterns of consumption that entirely contradict this logical explanation. If we were really interested in minimising the risk of tooth decay, we would brush our teeth after every meal, yet almost nobody does this. In fact, the times when people are most likely to clean their teeth occur *before* those moments when we are most frightened of the adverse social consequences of visible stains or bad breath.

When are you more likely to clean your teeth? Be honest. After eating ice cream, or when you're going on a date?* You might clean your teeth obsessively before giving a work presentation, or before

* Or, if you're married, how about before one of those rare moments in middle age when there is the faint possibility of sexual congress?

meeting someone for a romantic dinner. After eating a chocolate bar at home in the evening is perhaps not quite so likely. If you don't believe this, ask yourself one question: why is almost all toothpaste flavoured with mint? A recent trial proved that there were no dental-health benefits to the practice of flossing. I imagine that the manufacturers of dental floss were terrified by this finding, but they can relax – I confidently predict that this finding will have almost no effect on people's propensity to floss their teeth; they weren't really doing it for health reasons in the first place.†

Even stranger than our teeth-brushing behaviour is our preference for stripy toothpaste. When it first appeared, in a product called Stripe, it aroused a great deal of debate over how it was made. Many people dissected the empty container; others froze a full tube and then cut it open in a cross section.‡ What was strange was that nobody ever asked 'Why?' After all, the moment toothpaste enters your mouth, all the ingredients are mixed together, so what was the point of keeping them separate in the tube? There are two explanations: 1) simple childish novelty and 2) psycho-logic. Psychologically, the stripes serve as a signal: a claim that a toothpaste performed more than one function (fighting cavities, tackling infection and freshening breath) was thought to be more convincing if the toothpaste contained three visibly separate active ingredients. In general, people are impressed by any visible extra effort that goes

† I suspect people floss their teeth for the same reasons we love to put cotton buds in our ears – it just feels so damn good. I'm also a religious user of alcohol-based hand sanitisers, but if I am to be entirely honest with myself, I use them not to avoid infection, but for the delicious feeling you get when an alcohol-based gel evaporates on your hands.

‡ In fact there are two ways you can produce striped toothpaste, but as this is not a book about laminar flow, I won't go into the details here.

into a product: if you simply say 'this washing powder is better than our old powder', it is a hollow claim. However, if you replace the powder with a gel, a tablet or some other form, the cost and effort which have gone into the change make it more plausible to the purchaser there may have been some real innovation in the new contents.

The reason toothpaste is an especially interesting example is because, if an unconscious motivation happens to coincide with a rational explanation, we assume that it is the rational motive which drives the action.

Imagine you came home to find a dog turd on the floor of your kitchen; you would find it repellent, and would remove it immediately. Having disposed of it, you would wipe the floor with water and detergent, and if I asked you why you were doing those things, you would answer 'because it's unhygienic, of course; it's a source of germs'. But here's the thing; an early Victorian would have experienced exactly the same emotions and performed exactly the same actions, but they didn't know about germs. They were, technically speaking, irrational in their dislike of faeces, which was 'purely emotional'. Nowadays, if someone started flinging faeces around, we would describe him as a public-health hazard, while in the eighteenth century they would have called the practice 'ungodly' and in the fifteenth they might have burned him at the stake. So the dislike of faeces was not originally based on sound reasoning – it was rather a sound instinct the reason for which had not yet been discovered.

1.6 THE RIGHT THING FOR THE WRONG REASON

For the half-million or so years before medical bacteriology arose as a study in the 1870s, evolution had provided us with an emotional solution to a rational problem. You would be more likely to survive and reproduce if you had a strong aversion to poo, and so almost all of us are descended from people who disliked it. What's interesting is that we adopted the behaviour many thousands of years *before* we knew the reasons for it.

There is a good reason why evolution worked this way. Instincts are heritable, whereas reasons have to be taught; what is important is how you behave, not knowing why you do. As Nassim Nicholas Taleb remarks, 'There is no such thing as a rational or irrational belief – there is only rational or irrational behaviour.' And the best way for evolution to encourage or prevent a behaviour is to attach an emotion to it. Sometimes the emotion is not appropriate – for instance, there is no reason for Brits to be afraid of spiders, since there are no poisonous spiders in the UK – but it's still there, just in case. And why take the risk? Other than in a few specialist jobs in zoos, there's not much to be gained from not being afraid of spiders. So, as with tooth brushing, behaviours which have a rationally beneficial outcome do not have to be driven by a rational motivation. Cleaning our teeth is good for dental health even if we do it for reasons of vanity. As far as evolution is concerned, if a behaviour is beneficial, we can attach any reason to it that we like.

You don't need reasons to be rational.

History books are full of examples of public health or social benefits that have been driven by spiritual rather than material reasons.* Strict dietary law, in both Islam and Judaism, is a good example – and has a further benefit in the shape of social cohesion, as it forces people to eat together.

Additionally, while the rule against eating pigs may seem superstitious, as the anthropologist Richard Redding explains, keeping chickens rather than pigs has several key advantages. 'First, they are a more efficient source of protein than pigs; chickens require 3,500 litres of water to produce one kilo of meat, pigs require 6,000. Secondly, chickens produce eggs, an important secondary product which pigs do not offer. Third, chickens are much smaller and can thus be consumed within 24 hours; this eliminates the problem of preserving large quantities of meat in a hot climate. Finally, chickens could be farmed by nomads. While neither chickens nor pigs can be herded in the same way as cattle, chickens are small enough to be transported.' You can also add the risk of infection to this list; although in Judaism the prohibition against eating pork is described as *chok*, meaning a rule for which there is no rationale, pigs can spread diseases, and pig farming may pass them on to humans.

Similarly, Islam requires that the dead are buried as soon as possible after death, in order to 'reduce the suffering of the deceased in the afterlife and to return them to Allah'. As a result, throughout the Gallipoli campaign in 1915,† Muslims went to great lengths to bury their dead; by contrast, allied bodies often lay on the battlefield

* My thanks to Wing Commander Keith Dear for supplying two examples: fear of revenants and the accidental contribution this made to public health is described in Helga Nowotny's *The Cunning of Uncertainty* (2016).

† And no doubt in many battles before.

for days before they were collected. The outcome was further casualties from disease for the allies, and comparatively lower levels of disease among their opponents. Scientifically unverified beliefs about burial norms drove rational and life-saving behaviour.

Similarly, if asked why it was a good idea to create a space outside a town for the burial of the dead, a modern commentator might point to the risk of infection or pollution of the water supply. However, as I said above, we have only known about germs for a little over a hundred years, so why did towns build cemeteries away from their settlements long before this? Again it was an instinctive behaviour enshrined in a spiritual belief. In the Middle Ages, Europeans moved cemeteries from inside their fortifications to outside because of a fear that the souls of the bodies of the dead might return to haunt the living. The incidental result of this fear of 'revenants' was improved hygiene and protection from disease.

In trying to encourage rational behaviour, don't confine yourself to rational arguments.

Reason, and the naïve assumption that people understand the reasons for their own behaviour, would both provide very misleading explanations for the use of toothpaste. If you asked people why they cleaned their teeth, they would talk about dental health and avoiding trips to the dentist, probably without mentioning fresh breath and social confidence. A rational person would therefore suggest that people are motivated to practise health behaviours on the basis of their benefits; however, in reality, we probably mostly perform the healthy behaviour of cleaning our teeth for reasons altogether peripheral to the health benefit. My own view? Who cares why people clean their teeth, as long as they do it? Who cares why people recycle, as long as they do it? And who cares why people don't drink-drive, as long as they *don't* do it?

If you confine yourself to using rational arguments to encourage rational behaviour, you will be using only a tiny proportion of the tools in your armoury. Logic demands a direct connection between reason and action, but psycho-logic doesn't. This is important, because it means that, if we wish people to behave in an environmentally conscious way, there are other tools we can use other than an appeal to reason or duty. Similarly, if we wish to discourage people from drink-driving, we do not have to rely solely on rational arguments; if that approach does not work – and often it doesn't – there is a whole other set of emotional levers we can pull to achieve the same effect. Just ask the 1920s ad industry.

Believe it or not, the phrase 'Often a bridesmaid, never a bride' has its origins in an advert for Listerine – here was a hygiene product being sold not on medical benefits but on the fear of social and sexual rejection. 'Edna's case was a really pathetic one ... But that's the cruel thing about halitosis.' Similarly, a 1930s advert for Lifebuoy soap was headlined 'Why I cried after the party' – another product promoted on its romantic rather than its physiological benefits. Colgate's promise about 'the ring of confidence' was ingenious because it was ambiguous: it allowed the brand to talk about the confidence you would feel taking your children to the dentist, but also to imply the emotional confidence the product conferred on the user in a meeting or social situation.

Consumer behaviour, and advertisers' attempts to manipulate it, can be viewed as an immense social experiment, with considerable power to reveal the truth about what people want and what drives them. What people do with their own money (their 'revealed preferences') is generally a better guide to what they really want than their own reported wants and needs.‡

Had Darwin waited a hundred and fifty years or so, he could have saved himself a great deal of trouble and seasickness in uncovering our primate ancestry by travelling from Down House to his (and my) local Sainsbury's supermarket in Otford in Kent. There he could have learned from point-of-sale data that, over 30,000 items on the shelves, the single item most frequently purchased, as by all grocery shoppers in Britain, is ... a banana.

‡ In some ways, we need markets simply because prices are the only reliable means of getting consumers to tell the truth about what they want.

1.7 HOW YOU ASK THE QUESTION AFFECTS THE ANSWER

A few years ago, I was called by someone who was responsible for a programme to install smoke detectors in at-risk American housing. They had a problem: people were happy to receive a free smoke detector, but balked at having more than one installed. For instance, they might accept one in the entrance hall but decline one in a child's bedroom. I am sure that in the longer term there is a design solution to this problem – making smoke detection integral in light bulbs or lighting fixtures, for instance. However, my immediate suggestion was to borrow an approach from restaurant waiters and get people to accept three or four.

One of the great contributors to the profits of high-end restaurants is the fact that bottled water comes in two types, enabling waiters to ask 'still or sparkling?', making it rather difficult to say 'just tap'. I had the idea of turning up at an apartment with five smoke detectors; the fire officer was to casually carry in all five, before saying, 'I think we can make do with three here ... How many would you like, three or four?' We are highly social creatures and just as we find it very difficult to answer the question 'still or sparkling?' with 'tap', it is also difficult to answer the question about 'three or four' smoke detectors with with 'one'. As Nassim Nicholas Taleb remarks, 'the way a question is phrased is itself information'.

1.8 'A CHANGE IN PERSPECTIVE IS WORTH 80 IQ POINTS'

So said Alan Kay, one of the pioneers of computer graphics. It is, perhaps, the best defence of creativity in ten words or fewer. I suspect, too, that the opposite is also true: that an inability to change perspective is equivalent to a loss of intelligence.*

I was once walking down a suburban street in Wallingford, Pennsylvania. In American suburbia, there are no hedges to obscure the houses – a whitewashed fence about two feet high is all that marks property boundaries. So I was slightly alarmed when a large unleashed dog lurched towards me across one of the lawns, barking loudly. Clearly he was not going to have much difficulty clearing the little fence, after which he would be free to tear me to shreds. My companion, however, seemed unperturbed and sure enough, about two feet before the fence, the dog skidded to a halt on the lawn and continued its furious barking. As my friend knew, the dog was fitted with a collar that would detect the presence of a wire buried beneath the lawn boundary and administer an electric shock to the dog if it came too close.

* We all know people, I suspect, who though highly intelligent are insufferably inflexible in their approach to life.

Although the fence was only two feet high, the dog was terrified to approach it.

A similar constraint also applies to decision-making in business and government. There is a narrow and tightly limited area within which economic theory allows people to act. Once they reach the edges of that area, they freeze, rather like the dog. In some influential parts of business and government, economic logic has become a limiting creed rather than a methodological tool. As Sir Christopher Llewellyn Smith, the former director of CERN, remarked after he had been tasked with changing the patterns of energy consumption in Britain, 'When I ask an economist, the answer always boils down to just bribing people.'

Logic should be a tool, not a rule.

At its worst, neo-liberalism takes a dynamic system like free market capitalism, which is capable of spectacular creativity and ingenuity, and reduces it to a boring exercise in 'how we can buy these widgets 10 per cent cheaper'. It has also propelled a narrow-minded technocratic caste into power, who achieve the appearance of expert certainty by ignoring large parts of what makes markets so interesting. The psychological complexity of human behaviour is reduced to a narrow set of assumptions about what people want, which means they design a world for logical rather than psycho-logical people. And so we have faster trains with uncomfortable seats departing from stark, modernist stations, whereas our unconscious may well prefer the opposite: slower trains with comfortable seats departing from ornate stations.

This is not the fault of the markets; it is the fault of the people who have hijacked the definition of what markets should do. Strangely, as we have gained access to more information, data, processing power and better communications, we may also be losing the ability to see things in more than one way; the more data we have, the less room there is for things that can't easily be used in computation. Far from reducing our problems, technology may have equipped us with a rational straitjacket that limits our freedom to solve them.

Sometimes we grossly overrate 'the pursuit of reason', while at other times, we misdeploy it. Reasoning is a priceless tool for evaluating solutions, and essential if you wish to defend them, but it does not always do a very good job of finding those solutions in the first place. Maths, for instance, has the power to mislead as well as to illuminate. The inherent flaws of mathematical models are well understood by good mathematicians, physicists and statisticians, but very badly understood by those who are merely competent.†

Whenever I speak to a very good mathematician, one of the first things I notice is they are often sceptical about the tools which other mathematicians are most enthused about. A typical phrase might be: 'Yeah you could do a regression analysis, but the result is usually bollocks.' An attendant problem is that people who are not skilled at mathematics tend to view the output of second-rate mathematicians with an high level of credulity, and attach almost mystical significance to their findings. Bad maths is the palmistry of the twenty-first century.

Yet bad maths can lead to collective insanity, and it is far easier to be massively wrong mathematically than most people realise – a single dud data point or false assumption can lead to results that are wrong by many orders of magnitude.

In 1999 a British solicitor named Sally Clark was convicted for the murder of her two infant sons. Both had died of what was assumed to be cot death (or sudden infant death syndrome), a little over a year apart. After the death of the second child, suspicions were raised and she was prosecuted for murder. In a now-discredited piece of statistical evidence, the paediatrician Professor Sir Roy Meadow gave evidence at her trial that the chance of both deaths arising from natural causes was one in 73 million, or equivalent to

† Seriously good mathematicians are, as you would expect, much rarer than merely capable ones.

'four different horses winning the Grand National in consecutive years at odds of 80 to 1'. There are 700,000 live births in the UK every year, and there's a one in 8,453 chance of a child from a middle-class, non-smoking home dying from SIDS; Meadow had multiplied the two numbers together to get odds of one in 73 million, and claimed that you would only expect a double cot death in a family once in a hundred years.

One medical expert for the defence later described the figure as a gross statistical over-simplification, but it stuck – and there was a clear implication that there was only a statistically tiny chance that Clark was innocent. In a courtroom packed with scientists and lawyers, no one sought to rule this figure of one in 73 million inadmissible, but let's look at just how wrong it is. Firstly, it assumed odds of one in 8,453, taken from a single, selective source – whereas a more accurate figure might be closer to one in 1,500 or so. It also made no allowance for the fact that both victims were male, which reduces the odds further. Worst of all, it made no allowance for the possibility that a common combination of genes or an environmental factor – such as an aspect of the house in which they both died that was common to both tragedies – might have played a part. Genetic factors are believed to be involved in SIDS – it may run in families, making a double incidence much more likely.

As the journalist Tom Utley pointed out in the *Daily Telegraph*, he himself, among a circle of perhaps 10,000 acquaintances, knew two people who had each innocently lost two infants for inexplicable reasons, so the likelihood that this was as rare an occurrence as Professor Meadow had suggested seemed implausible. Although Sally Clark may have been unlucky in that on both occasions she was alone at home with her child and had no witnesses to defend her, if you make the corrections above you might reasonably expect double deaths to occur by chance several times a year in the UK – and thus it seemed less likely that Clark was guilty.

This, however, is still wrong. If you wish to prove the murderous intent of Sally Clark, it is not enough to prove that the cot-death theory is improbable. To do this is to fall victim to what is known as 'the prosecutor's fallacy', where the prosecution can imply that a similarity between the perpetrator and the accused carries more statistical weight than it deserves. (For instance, it may seem conclusive to suggest that a DNA marker shared by the perpetrator and a suspect is possessed by only one in 20,000 people, but if the suspect had been identified by trawling through a DNA database of 60,000 people, you would expect to find three people who exhibited this property, of whom at least two would be completely innocent.)

In Sally Clark's case, it is not enough to prove that double cot death is unlikely: you also have to prove that it is more unlikely than double infanticide. With the accurate statistical comparison, where you compare the relative likelihood of double cot death or double infanticide, the implied odds of her innocence fall from one in 73 million to perhaps two in three. She might still be guilty, but there is now more than enough reasonable doubt on which to acquit. Indeed, the most likely explanation is that she is innocent.

Notice, though, how just a few wrong assumptions in statistics, when compounded, can lead to an intelligent man being wrong by a factor of about 100,000,000 – tarot cards are rarely this dangerous. This miscarriage of justice led Professor Peter Green, as President of the Royal Statistical Society, to write to the Lord Chancellor, Britain's senior legal authority, pointing out the fallaciousness of Meadow's reasoning and offering advice on the better use of statistics in legal cases. However, the problem will never go away, because the number of people who think they understand statistics dangerously dwarfs those who actually do, and maths can cause fundamental problems when badly used.

To put it crudely, when you multiply bullshit with bullshit, you don't get a bit more bullshit – you get bullshit squared.

One thing this means is that everyone should know at least one seriously good mathematician; when you meet them, it is usually a revelation. I am proud to have met Ole Peters, a tremendous German physicist attached to the Santa Fe Institute and the London Mathematical Laboratory, in the last year. He recently co-authored a paper[§] pointing out that a huge number of theoretical findings in economics were based on a logical-sounding-but-entirely-erroneous assumption about statistical mechanics. The assumption was that, if you wished to know whether a bet was a good idea, you could simply imagine making it a thousand times simultaneously, add up the net winnings, and subtract the losses; if the overall outcome is positive, you should then make that bet as many times as you can.

So a bet costing £5 which has a 50 per cent chance of paying out £12 (including the return of your stake) is a good bet. You will win on average £1 every time you play, so you *should* play it a lot. Half the time you lose £5 and half the time you win £7. If a thousand people play the game just once, they will collectively end up with a net gain of £1,000. And if one person plays the game 1,000 times, he would expect to end up around £1,000 richer too – the parallel and series outcomes are the same. Unfortunately this principle applies only under certain conditions, and real life is not one of them. It assumes that each gamble is independent of your past performance, but in real life, your ability to bet is contingent on the success of bets you have made in the past.

Let's try a different kind of bet – one where you put in a £100 stake, and if you throw heads, your wealth increases by 50 per cent, but if you throw tails it falls by 40 per cent. How often would you want to toss the coin? Quite a lot, I suspect. After all, it's simple, right? All you have to do to calculate the expectation value over 1,000 throws is imagine 1,000 people taking this bet once simultaneously and average the outcome, like we did last time. If, on average, the group is better off, it therefore represents a positive expectation. But apparently not.

Let's look at it in parallel. If a thousand people all took this bet once, starting with £100 each (meaning a total of £100,000), typically 500 people would end up with £150 and 500 people would end up with

§ Co-authored with Murray Gell-Mann, who I can safely say is quite a good physicist, too, what with him getting a Nobel Prize and discovering quarks and all that. 'Evaluating Gambles Using Dynamics', *Chaos* (February 2016).

£60. That's £75,000 + £30,000 or £105,000, a net 5 per cent return. If someone asked me how often would I like them to toss the coin under those conditions, and how much would I like to put in, I would say: 'All the money I have, and throw the coin as fast as you can. I'm off to Mauritius with the winnings.' However, in this case the parallel average tells you nothing about the series expectation.

Put in mathematical language, an ensemble perspective is not the same as a time-series perspective. If you take this bet repeatedly, by far the most likely outcome is that you will end up skint. A million people all taking the bet repeatedly will *collectively* end up richer, but only because the richest 0.1 per cent will be multi-billionaires: the great majority of the players will lose. If you don't believe me, let's imagine four people toss the coin just twice. There are four possible outcomes: HH, HT, TH or TT, all of equal likelihood. So let's imagine that each of the four people starts with $100 and throws a different combination of heads and tails:

HH
HT
TH
TT

The returns on these four are £225, £90, £90 and £36. There are two ways of looking at this. One is to say, 'What a fabulous return: our collective net wealth has grown over 10 per cent, from £400 to £441, so we must all be winning.' The more pessimistic viewpoint is to say, 'Sure, but most of you are now poorer than when you started, and one of you is seriously broke. In fact, the person with £36 needs to throw three heads in a row just to recover his original stake.'

This distinction had never occurred to me, but it also seems to have escaped the attention of most of the economics profession, too. And it's a finding that has great implications for the behavioural sciences, because it suggests that many supposed biases which economists wish to correct may not be biases at all – they may simply arise from the fact that a decision which seems irrational when viewed through an ensemble perspective is rational when viewed through the correct time-series perspective, which is how real life is actually lived; what happens on average when a thousand people do something once is not a clue to what will happen when one person does something a thousand times. In this, it seems, evolved human instinct may be a much better at statistics than modern economists.¶ To explain this distinction using an extreme analogy, if you offered ten people £10m to play Russian roulette once, two or three people might be interested, but no one would accept £100m to play ten times in a row.

Talking to Ole Peters, I realised that the problem went wider than that – nearly all pricing models assume that ten people paying for something once is the same as one person paying for something ten times, but this is obviously not the case. Ten people who each order ten things every year from Amazon will probably not begrudge paying a few dollars for delivery each time, while one person who buys 100 things from Amazon every year is going to look at his annual expenditure on shipping and decide, 'Hmm, time to rediscover Walmart.'**

One of our clients at Ogilvy is an airline. I constantly remind them that asking four businessmen to pay £26 each to check in one piece of luggage is not the same as asking a married father of two†† to pay £104 to check in his family's luggage. While £26 is a reasonable fee for a service, £104 is a rip-off. The way luggage pricing should work is something like this: £26 for one case, £35 for up to three. There is, after all, a reason why commuters are offered season tickets – commuting is not commutative, so 100 people will pay more to make a journey once than one person will pay to make it 100 times. In the same way, the time-saving model used to justify the UK's current investment in the High Speed 2 rail network assumes that 40 people saving an hour ten times a year is the same as one commuter saving an hour 400 times a year. This is obviously nonsense; the first is a convenience, while the second is a life-changer.

¶ Presumably, the people who thought like economists all died out.

** This explains why Amazon Prime needs to exist. Without it, Amazon cannot have regular customers.

†† Me, dammit!

1.9 BE CAREFUL WITH MATHS: OR WHY THE NEED TO LOOK RATIONAL CAN MAKE YOU ACT DUMB

I would rather run a business with no mathematicians than with second-rate mathematicians. Remember that every time you average, add or multiply something, you are losing information. Remember also that a single rogue outlier can lead to an extraordinary distortion of reality – just as Bill Gates can walk into a football stadium and raise the average level of wealth of everyone in it by $1m.

The advertising agency I work for once sent out postal solicitations for a charity client, and we noticed that one creative treatment significantly outperformed the other in its net return. Since there was not much difference between the treatments, the scale of the difference in results surprised us. When we investigated, we found that one person had replied with a cheque for £50,000.*

Let's look at another example of how one rogue piece of data – a single outlier – can lead to insane conclusions when not understood in the proper context. I have a card which I use to pay for my car's fuel, and every time I fill up, I record the car's mileage on the payment terminal. After a year, the fuel-card company started putting

* A recent lottery winner?

my mileage per gallon on my monthly statements – a lovely idea, except it started to drive me insane, because every month my car became less economical. I was puzzling over this for ages, agonising about fuel leaks and even wondering if someone was pilfering from my tank.

But then I remembered: soon after my company had given me the fuel card, I had forgotten to use it once, and had instead paid with an ordinary credit card. This meant that, according to the data held by the fuel-card company, in one period I had driven the distance of two tankfuls with one tankful of fuel. Because this anomaly was still sitting in the database, every subsequent month I drove was making my fuel-economy stats look worse as I was regressing to the mean – one anomalous data point made all the rest misleading.

But let me come back to my previous point. In maths, 10 x 1 is always the same as 1 x 10, but in real life, it rarely is. You can trick ten people once, but it's much harder to trick one person ten times.† But how many other things are predicated on such assumptions? Imagine for a moment a parallel universe in which shops had not been invented, and where all commerce took place online. This may seem like a fantastical notion, but it more or less describes rural America a hundred years ago. In 1919 the catalogues produced by Sears, Roebuck and Company and Montgomery Ward were, for the 52 per cent of Americans in rural areas, the principal means of buying anything remotely exotic. In that year, Americans spent over $500 million on mail-order purchases, half of which were through the two companies.

Yet in 1925, Sears opened its first bricks-and-mortar shop. By 1929, the companies had opened a further 800 between them – perhaps Amazon's purchase of Whole Foods Market is history repeating

† This is why conmen tend to base themselves in cities, on racecourses and in other places that offer a reliable supply of gullible victims.

itself.[‡] I could go on endlessly about the psychological factors at play here, but let's go back to the lazy assumption that 1 x 10 is the same as 10 x 1, which is also relevant.[§] Online shopping is a very good way for ten people to buy one thing, but it is not a good way for one person to buy ten things. Try and buy ten different things simultaneously online[¶] and it turns chaotic. Items arrive on four separate days, vans appear at your house at different times and one delivery always fails.[**] By contrast, the great thing about Walmart, which investors tend to overlook, is that people turn up, buy 47 different things and then transport them home at their own expense. Amazon can be a very big business selling one thing to 47 people, but if it can't sell 47 things to one person, there's a ceiling to how large it can be.

Many other mathematical models involving humans make the mistake of assuming that 10 x 1 = 1 x 10. For instance, our tax system assumes that ten people who earn £70,000 for one year of their life should be taxed the same amount as one person who earns £70,000 for ten consecutive years, yet I have never heard anyone question this – is it another example of bad maths?

Recently I was involved in a discussion of train overcrowding.[††] Again, metrics do not distinguish between ten people who have to stand 10 per cent of the time and one person who has to stand all the time, but these two things aren't the same at all. If I am an occasional traveller and find myself standing once a month, so be

‡ Certainly we Brits might feel smug about it since, by moving into physical retailing, Amazon seems to have discovered that Argos had it right all along: a physical presence still counts.

§ I call it 'Sutherland's Law of Bad Maths'.

¶ As you may do before Christmas.

** Requiring you to drive to an industrial estate in Dartford on Christmas Eve, thus wiping out any putative time savings.

†† As you may have guessed, it's an obsession of mine.

it, but if I had paid £3,000 for a season ticket and didn't ever get a seat, I would feel robbed. Reframed this way, the problem becomes easier to solve. Why not run two trains in each direction each day exclusively for season ticket holders, or give them the right to sit in first class when standard class is full?‡‡ Or better still, expand first class and allow all season ticket holders to sit in it. You haven't solved the problem of train overcrowding, but you have solved it for those who are worst affected, which is what really matters.

‡‡ I do not use a season ticket, but I would find both these arrangements perfectly equitable – just as I would expect someone who regularly eats at a restaurant to be offered a better table.

1.10 RECRUITMENT AND BAD MATHS

Okay. So it might work for trains. But if I told you that you could use a similar insight to increase employment diversity, you probably wouldn't believe me. But again, 10 x 1 does not equal 1 x 10. Imagine you have ten roles to fill, and you ask ten colleagues to each hire one person. Obviously each person will try to recruit the best person they can find – that's the same as asking one person to choose the best ten hires he can find, right? Wrong. Anyone choosing a group of ten people will instinctively deploy a much wider variance than someone hiring one person. The reason for this is that with one person we look for conformity, but with ten people we look for complementarity.

If you were only allowed to eat one food, you might choose the potato. Barring a few vitamins and trace minerals, it contains all the essential amino acids you need to build proteins, repair cells and fight diseases – eating just five a day would support you for weeks. However, if you were told you could only eat ten foods for the rest of your life, you would not choose ten different types of potato. In fact, you may not choose potatoes at all – you would probably choose something more varied.

The same applies to hiring – we are much more likely to take risks when hiring ten people than when hiring one. If you hire ten people, you might expect one or two of them not to work out: you won't risk your reputation if a couple of them leave after two years, or if one starts stealing staplers and photocopying his bottom at the Christmas party. But if you hire one person and they go rogue, you have visibly failed. So individuals who are hiring individuals may be needlessly risk averse; they are hiring potatoes.

When hiring, we should understand that unconscious motivation and rational good sense overlap, but they do not completely coincide. A person engaged in recruitment may think they are trying to hire the best person for the job, but their subconscious motivation is subtly different. Yes, they want to hire a candidate who is likely to be good, but they are also frightened of hiring someone who might turn out to be bad – a low variance will be as appealing to them as a high average performance. If you want low variance, it pays to hire conventionally and adhere to the status quo, while people hiring a *group* of employees are much more likely to take a risk on some less conventional candidates.

We can see this diversity mechanism clearly in house hunting. If I were to give you a budget to choose your perfect house, you would have a clear idea of what to buy, but it would typically be a bit boring. That's because when you have one house, it cannot be too weak in any one dimension: it cannot be too small, too far from work, too noisy or too weird, so you'll opt for a conventional house. On the other hand, if I were to double your budget and tell you to buy *two* houses, your pattern of decision-making would change. You would now be looking to buy two significantly different properties with complementary strengths – perhaps a flat in the city and a house in the countryside.

If you are choosing a parliamentary candidate, the safe option is to pick a vapid-but-presentable PPE graduate, yet no one choosing ten candidates would pick *ten of those* – they'd throw in a few wild cards.* Cecil 'Bertie' Blatch clearly understood this when, as President of the

* Perhaps someone who'd had a proper job, someone from a poorer background and someone with a science degree.

Finchley and Friern Barnet Conservative Association, he decided to 'lose' a couple of a more conventional candidate's votes to give Margaret Thatcher her first winnable seat. He wasn't cheating; he was correcting a mental bias. Once you understand this, the potential to increase the level of diversity in employment, education or politics, without imposing quotas, rises as people are chosen in batches.†

Everyone worries about declining social mobility, rising inequality and the hideous homogeneity of politicians, yet it is possible these have arisen from well-meaning attempts to make the world fairer. The quandary is that you can either create a fairer, more equitable society, with opportunities for all but where luck plays a significant role, or you can create a society which maintains the illusion of complete and non-random fairness, yet where opportunities are open to only a few – the problem is that when 'the rules are the same for everyone' the same boring bastards win every time. The idea that you should therefore try making your recruitment system less fair outrages people when I suggest it, but it is worth remembering that there is an inevitable trade-off between fairness and variety. By applying identical criteria to everyone in the name of fairness, you end up recruiting identical people.‡

At Ogilvy we now recruit creative talent through an internship scheme called 'The Pipe'. Applicants don't have to be graduates; they don't have to be young; they don't have to have any

† I know this from personal experience. Years after I was first hired, someone involved in the selection process revealed that I would never have been offered a job had they been recruiting one person at a time, but because they had four vacancies they decided 'to take a punt on the weirdo', or words to that effect.

‡ Should you at least offer an interview to someone with a rotten degree who was, say, the reigning under-25 UK backgammon champion? The 'fairness police' would say no, but personally I would see them every time.

qualifications at all – in fact, we recruit them blind for the first few stages. It is too early to make any definitive judgement about the programme's success, but the recruits seem every bit as interesting to chat to as the Oxford graduates – and more interesting, in some cases.[§] Within months of our first intake joining, several of them had won an award at the Cannes Festival of Creativity for an advertising idea for Hellmann's Mayonnaise – something some people spend a lifetime in the industry without achieving.

Remember, anyone can easily build a career on a single eccentric talent, if it is cunningly deployed. As I always advise young people, 'Find one or two things your boss is rubbish at and be quite good at them.' Complementary talent is far more valuable than conformist talent.

§ One is a former mixologist and another a poet.

1.11 BEWARE OF AVERAGES

When Lieutenant Gilbert S. Daniels, a physical anthropologist, was hired by the US military to design a better cockpit for high-speed aircraft in the early 1950s, the assumption he had to challenge was that you should design a cockpit for 'the average man'. The idea was that if you took an average of many pilots' bodily dimensions, you would have a template around which you could design a cockpit – with the instruments visible to most people, and with the controls within easy reach of all but the most unusual physical specimens.

However, Daniels already knew from his measurements of human hands that an average human hand is not a typical human hand, and in the same way he found that an average human body – that is one which is average in a range of dimensions – is surprisingly rare. When you designed a cockpit for an average man you were designing a cockpit not for everyone, but for a surprisingly rare, or even non-existent, body-type. Not a single pilot of the 4,000 measured was within the average range on all ten bodily measures.*

* Likewise, an attempt a few years earlier to find a perfect female form by averaging female body-types similarly foundered.

Don't design for average.

Metrics, and especially averages, encourage you to focus on the middle of a market, but innovation happens at the extremes. You are more likely to come up with a good idea focusing on one outlier than on ten average users. We were discussing this recently in a meeting when a round of sandwiches arrived. 'This proves my point exactly,' I said, pointing at the food. The sandwich was not invented by an average eater. The Earl of Sandwich was an obsessive gambler, and demanded food in a form that would not require him to leave the card table while he ate.

Weird consumers drive more innovation than normal ones. By contrast, it is perfectly possible that conventional market research has, over the past fifty years, killed more good ideas than it has spawned, by obsessing with a false idea of representativeness.

1.12 WHAT GETS MISMEASURED GETS MISMANAGED

The above findings have important implications for metrics. Businesses love them, because they make it easy to compare and manage things. It's true that 'what gets measured gets managed', but the concomitant truth is 'what gets mismeasured gets mismanaged'. One great problem with metrics is that they destroy diversity because they force everybody to pursue the same narrow goal, often in the same narrow way, or to make choices using the exact same criteria.

I have never seen any evidence that academic success accurately predicts workplace success.* The eminent obstetrician and fertility specialist Professor Lord Winston does not seek out academic high-flyers to work with in medicine, even though he could have the pick of the crop. However, it is now common practice in British firms only to interview people with an upper second-class degree or above, a criterion that is applied with no evidence but simply because it seems logical. If you need to select from a list of graduate

* In medicine it used to be said – only partly in jest – that the opposite rule applied: that those students with second-class degrees went on to become the best doctors, and people with third-class degrees went on to become the richest doctors.

applicants, using their undergraduate performance as a filter seems to make sense, but without more evidence it is absurd – and when practised on a wide scale, it's a disgraceful waste of talent.

It wouldn't matter much if only Goldman Sachs, say, or a few elite institutions used this criterion, but when everyone else copies the same approach, it is ludicrous. Since almost half of graduates should by definition fall below this hurdle, it will either result in thousands of people spending three years at university for no benefit or to grade inflation in universities, with degree classes becoming meaningless.† This is another example of people not using reason to make better decisions, but simply for the appearance of being reasonable.

As any game theorist knows, there is a virtue to making slightly random decisions that do not conform to established rules. In a competitive setting such as recruitment, an unconventional rule for spotting talent that nobody else uses may be far better than a 'better' rule which is in common use, because it will allow you to find talent that is undervalued by everyone else.

One of the other problems with a logically consistent system for hiring people is that ambitious middle-class people can exploit it by 'gaming the system'. Violin lessons – check, work placement at uncle's bank – check, charity work with the disadvantaged – check,‡ high GPA – check. By contrast, if you hire that brilliant backgammon player, you do know one thing. He's genuinely talented at something – and it's unlikely that his parents have spent a fortune on private backgammon lessons.

† I recently met someone who had a lower second-class degree in Mathematics from Cambridge who found it difficult to get a job interview. How can this be? What a nonsense!

‡ I sometimes wonder whether homeless shelters sometimes secretly wish all those aspiring Ivy League and Oxbridge applicants would just piss off home.

Real excellence can come in odd packaging. Nassim Nicholas Taleb applies this rule to choosing a doctor: you don't want the smooth, silver-haired patrician who looks straight out of central casting – you want his slightly overweight, less patrician but equally senior colleague in the ill-fitting suit. The former has become successful partly as a result of his appearance, the latter despite it.

1.13 BIASED ABOUT BIAS

Is some of what we think to be racial or gender bias merely bias in favour of the status quo? After all, the less eccentric your hire, the less blame you are exposed to if something goes wrong. The advertising industry at present is obsessed with gender ratios and ethnic composition; it is perfectly reasonable to look at these figures, but the industry seems completely blind to another bias, which is that it is extraordinarily prejudiced in favour of hiring physically attractive people.

Understanding what is going on subconsciously is vital here – if we assume that racial bias is the only subconscious bias, we may introduce misdirected measures. If, however, we consider the possibility that part of the issue regarding a lack of diversity arises from status quo bias, the solutions will be very different. I am not denying for a moment that unconscious racial bias exists, but I am suggesting that people may have an unhealthy predisposition to attribute all disparities of outcome and opportunity between different ethnic groups to it, whereas in reality many other forces may be at work. If you want to solve the problem, you have to understand 'the real why'.

Some evolutionary psychologists, most notably Robert Kurzban, believe that racial prejudice is a relatively weak force in human psychology since for most of evolutionary history we wouldn't have encountered people of a different ethnicity. We might, rather, be heavily predisposed to attaching 'outsider status' to people who speak with a different accent to us, since such distinctions would have been experienced more frequently. Coming from a country that historically was obsessed with accents,* this does seem worth investigating. Would a privately educated Nigerian be at a disadvantage in seeking a job in London in competition with a white Liverpudlian with a strong Scouse accent? I rather doubt it. Kurzban's work suggests that prejudice depends on context; we may be more prone to ascribe 'outsider status' to someone of a different ethnic background, but this does not mean that we cannot easily adopt anyone of any ethnic background into an 'in group' in a different setting.

I am slightly sceptical about using gender and ethnic quotas as a way of tackling prejudice, since they define diversity in narrow ways. For instance, it would be perfectly possible to improve your racial diversity figures by employing ten highly talented Nigerians. You might congratulate yourself for the admirable diversity shown by your firm, but what if you were to then find out that all ten came from the Igbo tribe and none come from the almost equally populous Yoruba tribe – would you bury this fact and remain smug about your newly diverse workforce, or would you ask whether your definition of diversity is a little heavy on weighting skin colour and a bit light on everything else?

A recent article in *Harvard Business Review* showed that apparent gender or racial bias may not only arise for the assumed reasons; other unconscious mechanisms may also be involved, depending on the context. It reported on three studies which examined what happens when you change the finalists for a job. The research revealed that: 'When there is only one woman, she does not stand a chance of being hired, but that changes dramatically when there is more than one. Each added woman in the pool does not increase the probability of hiring a woman, however – the difference

* And still is, though to a lesser extent.

The relationship between finalist pools and actual hiring decisions

According to one study of 598 finalists for university teaching positions

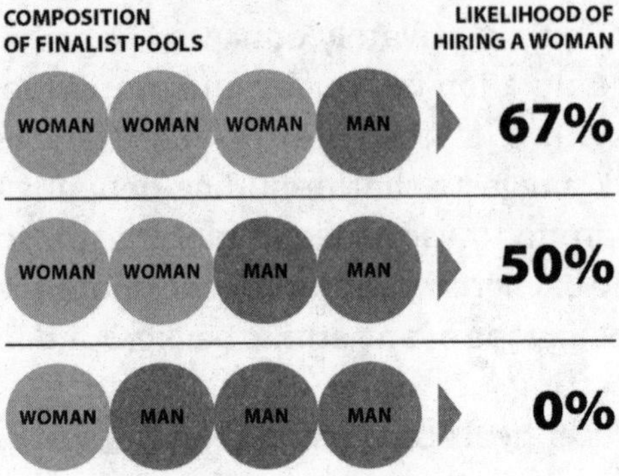

Are people biased against minorities – or are they biased against anyone in a minority of one?

between having one and two women seems to be what matters. There were similar results for race when we looked at a pool of four candidates.'

This suggests that the prejudice we apply against a lone black candidate or a lone female candidate might also apply to a lone 'anything' candidate.†

† Try going to your next job interview and performing magnificently in every way, but insisting on wearing a hat throughout. I'm willing to bet that you won't get the job. Unless, of course, you can persuade another candidate to do the same.

1.14 WE DON'T MAKE CHOICES AS RATIONALLY AS WE THINK

The context and order of choosing affects things in ways we would not consciously expect – not just in corporate decision-making and recruitment, but also in personal decisions. The psychologist and behavioural economist Dan Ariely was one of the first people to highlight the famous decoy effect* in the decision process – the phenomenon whereby consumers tend to have a specific change in preference between two options when also presented with a third option that is more desirable than one, but less desirable than the other.

A seminal example explored by Ariely is the *Economist* magazine's subscription offer. The middle option – where you are offered a print only subscription for $125 – is known as a decoy. No one – except perhaps someone who deeply loathes technology – would ever choose it, since for the same price you could get the full print and online subscription, but it does have a huge effect on behaviour. By creating a very easy 'no-brainer' decision, it encourages more people to take up the higher-value full subscription. In one experiment carried out by Ariely, it led 84 per cent of putative subscribers to choose the full-price, all-in subscription. However, when you remove the dummy option and only have the two

* Also known as the asymmetric dominance effect.

SUBSCRIPTIONS

Welcome to
The Economist Subscription Centre

Pick the type of subscription you want to buy or renew.

☐ **Economist.com subscription** – US $59.00
One-year subscription to Economist.com
Includes online access to all articles from *The Economist* since 1997.

☐ **Print subscription** – US $125.00
One-year subscription to the print edition of *The Economist*.

☐ **Print & web subscription** – US $125.00
One-year subscription to the print edition of *The Economist* and online access to all articles from *The Economist* since 1997.

Context is everything: strangely, the attractiveness of what we choose is affected by comparisons with what we reject. As one friend remarked, 'Everyone likes to go to a nightclub in the company of a friend who's slightly less attractive than them.'

sensible ones, preferences are reversed: 68 per cent chose the lower-priced, online-only subscription.

Estate agents sometimes exploit this effect by showing you a decoy house, to make it easier for you to choose one of the two houses they really want to sell you. They will typically show you a totally

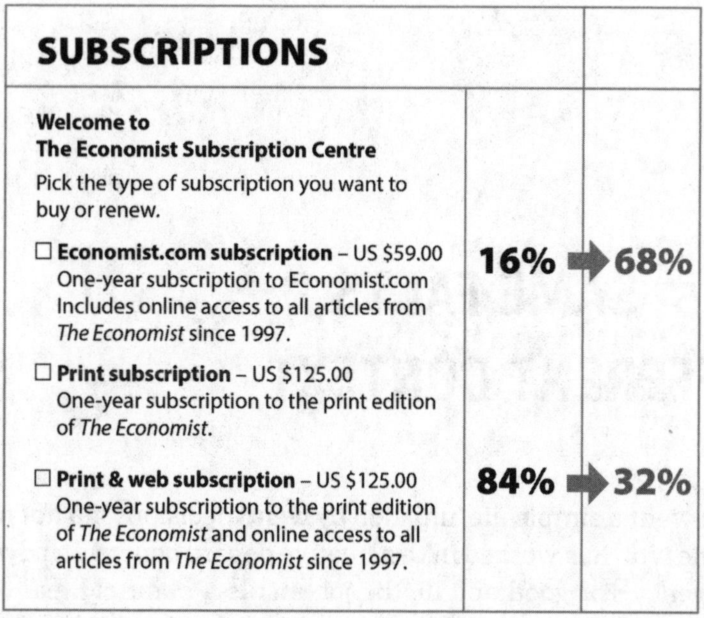
SUBSCRIPTIONS
Welcome to
The Economist Subscription Centre
Pick the type of subscription you want to buy or renew.
☐ Economist.com subscription – US $59.00
One-year subscription to Economist.com
Includes online access to all articles from The Economist since 1997.
☐ Print subscription – US $125.00
One-year subscription to the print edition of The Economist.
☐ Print & web subscription – US $125.00
One-year subscription to the print edition of The Economist and online access to all articles from The Economist since 1997.
16%
68%
84%
32%

inappropriate house and then two comparable houses, of which one is clearly better value than the other. The better value house is the one they want to sell you, while the other one is shown to you for the purpose of making the final house seem really good.

Here again you see examples of human action where the same behavioural quirks in a small and relatively inconsequential level of decision-making – choosing a holiday or a magazine subscription – are mirrored in bigger decisions. You might accept that a decoy or asymmetric dominance might affect your choice of magazine or holiday. But surely it wouldn't apply to something as momentous as buying a house or hiring staff? Sorry, but it does – the peculiarities of human decision-making seem to apply at all levels. It's one reason why I believe academics and policy-makers might benefit from paying more attention to consumer marketing. Tiny things that you discover when selling bars of chocolate can be relevant in how you encourage more consequential behaviour. Usually someone has often already found an answer to your problem – just in a different domain.

1.15 SAME FACTS, DIFFERENT CONTEXT

If you want a simple life, unladen by weird decisions, do not marry anyone who has worked in the creative department of an advertising agency. For good and ill, the job instils a paranoid fear of the obvious and fosters the urge to question every orthodoxy and to rail against every consensus. This becomes tiring – especially when the same wilfully perverse thinking is applied to everyday household decisions.

A few years ago, our family toaster was not only prone to producing alarming sparks and occasional outbursts of flames or smoke, but its slot was far too narrow, meaning that every slice of bread thicker than industrial white sliced was liable to get stuck between the toasting elements.* 'Why don't you buy one of those new wide-slot toasters?' my wife suggested. An hour or so later I returned home carrying a massive box and revealed to my wife that it contained, not a new toaster, but a bread-slicing machine. 'I rewrote the brief,' I declared proudly. 'We don't need a wider toaster. What we need is *narrower bread*!'

* The two problems were quite possibly connected – the sparks and flames came from lumps of wholemeal bread that had become trapped inside.

We tried this solution for a time, slicing bread thinly to fit in the narrow toaster. It wasn't altogether hopeless, but the bread slicer occupied about half the available work surface in the kitchen and generated a spectacular amount of crumbs. Then we had children, and the lethal revolving blade had to be kept out of reach of their small hands. Today the machine sits in a cupboard, while above it on the shelf is a heavily used, wide-slot toaster, just as my wife had originally suggested.

But …

The cupboard in which the bread machine sits is in the kitchen of our four-bedroom flat, which is on the second floor of a building constructed in around 1784. The house was built for the personal doctor of King George III by the architect Robert Adam, one of the titans of eighteenth-century British architecture. It sits in seven acres of communal grounds shaped by Capability Brown, the English landscape architect responsible for designing the gardens of Blenheim Palace and Highclere Castle.† And I got all this for free. I didn't get the apartment for free, obviously. It cost me £395,000 in 2001. Its market value now is perhaps £650,000 but, if you buy it, you get the architecture and the landscaping for free.‡ The building is Grade I listed, placing it in the top 2.5 per cent of the 375,000 listed buildings in England, almost half of which are churches, while many of the others are uninhabitable – Nelson's Column, for example, or the Royal Opera House.

So there are probably around 2,500 Grade I listed buildings in England in which you can actually live,§ and I pay nothing at all for the privilege. Whereas a painting by Picasso costs perhaps 100,000 times more than a picture bought from a Sunday exhibitor on the Bayswater Road, a house designed by Robert Adam costs

† The latter is now better known as Downton Abbey.

‡ As well as a free bread slicer left behind by the previous occupants.

§ Including Buckingham Palace.

If you want to buy really cheap art, buy architecture.

no more than an identically sized house by an unknown architect in the same area. Recently a flat by the modernist architects Maxwell Fry and Walter Gropius went on sale in Notting Hill; because it was in Notting Hill it was insanely expensive, but it was no more expensive than the banal apartments in the next-door house.

The reason I enjoy this spectacular architecture at no cost at all is because I deployed exactly the same perverse reasoning when buying a house as I did when buying the bread slicer: I rewrote the brief, and tried to make a decision while disposing of the usual assumptions. I wondered what most people do when they move house, aware that if I chose a house the way most people do, I would end up competing with a lot of people for the same houses. On the other hand, I knew that if I bought a house using wildly divergent criteria from everyone else, I should find a place that was relatively undervalued. In competitive markets, it pays to have (and to cultivate) eccentric tastes.

When most people buy a house, the order of search is as follows: 1) set a price band, 2) define location, 3) define number of bedrooms, 4) set other parameters – garden size, for example. Architectural quality comes low on the list – and is further devalued because it isn't quantifiable. If you can convince yourself to value something which other people don't, you can enjoy a fabulous house for much less.¶

I had decided before we moved that I wanted to live somewhere interesting, placing more emphasis on the architecture than on the precise location or the number of bedrooms. This eccentric approach certainly minimises status envy. Occasionally we visit insanely expensive houses owned by friends. 'What did you think?' my wife will ask as we drive home. 'Well it's certainly big,' I reply, 'but I couldn't help thinking the architecture was a bit rubbish.'

As I said, our apartment is on the second floor, and there is no lift.** But again, I decided to look at it differently. Not having a lift is good for you, because several times a day you get guaranteed exercise. In my mind, the flat no longer suffered from the absence of a lift – it was blessed with a free gym.

There are two lessons to be learned here. Firstly, it doesn't always pay to be logical if everyone else is also being logical. Logic may be a good way to defend and explain a decision, but it is not always a good way to reach one. This is because conventional logic is a straightforward mental process that is equally available to all and will therefore get you to the same place as everyone else. This isn't always bad – when you are buying mass-produced goods, such as toasters, it generally pays to cultivate mainstream tastes. But when choosing things in scarce supply†† it pays to be eccentric. The second interesting thing is that we have no real unitary measure of what is important and what is not – the same quality (such as not having a lift) can be seen as a curse or a blessing, depending on how you think of it. What you pay attention to, and how you frame it, inevitably affects your decision-making.

¶ American readers of this book might like to visit the website Wright On The Market, which lists the current Frank Lloyd Wright properties for sale.

** The apartment on the floor below would cost £200,000 more than ours, largely for this reason.

†† Such as property, beaches or spouses.

In making decisions, we should at times be wary of paying too much attention to numerical metrics. When buying a house, numbers (such as number of rooms, floor space or journey time to work) are easy to compare, and tend to monopolise our attention. Architectural quality does not have a numerical score, and tends to sink lower in our priorities as a result, but there is no reason to assume that something is more important just because it is numerically expressible.

1.16 SUCCESS IS RARELY SCIENTIFIC – EVEN IN SCIENCE

We often misuse our powers of reason, setting too low a bar in how we evaluate solutions, but too high a bar in our conditions for how we reach solutions. Reason is a wonderful evaluative tool, but we are treating it as though it were the only problem-solving tool – it isn't. If you look at the history of great inventions and discoveries, sequential deductive reasoning has contributed to relatively few of them. Graphene, one of the most important discoveries of the last 30 years, was discovered by the physicist Andre Geim in Manchester,* but he created the substance by messing around with pencils and Sellotape, equipment that any of us could have bought at a branch of Staples.

Geim says of his approach to science: 'I jump from one research subject to another every few years. I do not want to study the same stuff "from cradle to coffin", as some academics do. To be able to do this, we often carry out what I call "hit-and-run experiments", crazy ideas that should never work and, of course, they don't in most cases. However, sometimes we find a pearl ... This research style may sound appealing but it is very hard psychologically, mentally, physically, and in terms of research grants, too. But it is fun.'

* A discovery which won him a Nobel Prize.

For all we obsess about scientific methodology, Geim knows it is far more common for a mixture of luck, experimentation and instinctive guesswork to provide the decisive breakthrough; reason only comes into play afterwards. The bureaucrats to whom he must justify his activities, however, demand reasons right from the beginning to justify funding, but the idea that there is a robust scientific process that will reliably lead to progress seems unfounded.

Here is the brilliant American physicist Richard Feynman, in a Lecture in 1964, describing his method: 'In general, we look for a new law by the following process. First, we guess it ... Then we compute the consequences of the guess, to see what, if this law we guess is right, to see what it would imply and then we compare the computation results to ... experience, compare it directly with observations to see if it works ... In that simple statement is the key to science. It doesn't make any difference how beautiful your guess is, it doesn't matter how smart you are, who made the guess or what his name is ... If it disagrees with the experiment, it's wrong. That's all there is to it.'

A good guess which stands up to observation is still science. So is a lucky accident.

Business people and politicians do not quite understand this and tend to evaluate decisions by the rigour of the process that produces them, rather than by the rigour with which you evaluate their consequences. To them, the use of reason 'looks scientific', even if it is being used in the wrong place. After all, should we refuse to use antibiotics, X-rays, microwave ovens or pacemakers because the scientific discoveries which led to their creation were the product of lucky accidents?† You would have to be a deranged purist to adopt this view – and you would also end up hungry, bored and quite possibly dead. As with scientific progress, so too

† For more on this, see Paul Feyerabend's masterpiece *Against Method* (1975), or the writings of Sir Peter Medawar.

with business. The iPhone, perhaps the most successful product since the Ford Model T, was developed not in response to consumer demand or after iterative consultation with focus groups; it was the monomaniacal conception of one slightly deranged man.‡

And yet, in the search for public policy and business solutions, we are in the grip of an obsession with rational quantification. A nervous and bureaucratic culture is closed-mindedly attaching more importance to the purity of the methodology than to the possible value of the solution, which leads us to ignore possible solutions not because they have been proven to be wrong, but because they have not been reached through an approved process of reasoning.

A result of this is that business and politics have become far more boring and sensible than they need to be. Steve Jobs's valedictory injunction to students to 'stay hungry, stay foolish' probably contained more valuable advice than may be apparent at first glance. It is, after all, a distinguishing feature of entrepreneurs that, since they don't have to defend their reasoning every time they make a decision, they are free to experiment with solutions that are off-limits to others within a corporate or institutional setting.§

We approve reasonable things too quickly, while counterintuitive ideas are frequently treated with suspicion. Suggest cutting the price of a failing product, and your boringly rational suggestion will be approved without question, but suggest renaming it and you'll be put through gruelling PowerPoint presentations, research groups,

‡ And one with a highly unusual phobia; Steve Jobs had koumpounophobia, or a fear of buttons. See page 217.

§ When IBM created a PC division, it placed it in Florida, a whole seaboard's length away from their head office in New York. It did this to prevent managerialists from stifling new ideas at birth, and to provide experimental space for what T.J. Watson called 'wild ducks'.

multivariate analysis and God knows what else¶ – and all because your idea isn't conventionally logical. However, most valuable discoveries don't make sense at first; if they did, somebody would have discovered them already. And ideas which people hate may be more powerful than those that people like, the popular and obvious ideas having all been tried already.

We should test counterintuitive things – because no one else will.

¶ The process will take months of (and off) your life.

1.17 THE VIEW BACK DOWN THE MOUNTAIN: THE REASONS WE SUPPLY FOR OUR EXPERIMENTAL SUCCESSES

Imagine you are climbing a large mountain that has never been climbed before. From the bottom, it is impossible to tell which slopes are passable, because much of the terrain is hidden behind the lower foothills. Your climb involves a great deal of trial and error: routes are tried and abandoned; there is frequent backtracking and traversing. Many of the decisions you take may be based on little other than instinct or good fortune. But eventually you do make it to the summit, and once you are there, the ideal route is apparent. You can look down and see what would have been the best path to have taken, and that now becomes 'the standard route'. When you describe the route you took to your mountaineering friends, you pretend it was the route you took all along: with the benefit of hindsight, you declare that you simply chose that route through good judgement.

Is this a lie? Well, yes and no.* It may be that, in the course of your climb, you did end up at various times covering most or all of the optimal route.† What you say is also true in so far as it confirms that there is a navigable pathway to the top, which you did not know

* Well, mostly yes.

† Though it is also possible that your real route and the optimal route did not intersect at all.

for sure when you first attempted the climb. And the route you describe *does* exist, so in that sense your description of the climb is perfectly accurate. However, in one respect it is a monstrous lie, because it completely misrepresents the process by which you progressed to the top. It pays an undue tribute to rational decision-making, optimisation and sequential logic – a tribute that really should be laid at the altar of trial and error, good instincts, and luck.‡

As I write this, a TV detective drama is playing on the television, in which exactly the same sort of 'selective editing' is used to describe the apprehension of the murderer. The convention of detective drama is that you only refer to the information that has a bearing on the apprehension of the criminal, while in detective fiction, one or two red-herrings are allowed: however, in neither are the long hours of wasted legwork and time spent in pursuit of unrewarding lines of enquiry ever shown. As Alfred Hitchcock once said, 'drama is just real life with the boring bits edited out'.

We constantly rewrite the past to form a narrative which cuts out the non-critical points – and which replaces luck and random experimentation with conscious intent. For instance, a friend of mine once mentioned that he had been attracted to buy his current home partly because it was close to an excellent restaurant, forgetting that the establishment opened after he had moved in. In reality, almost everything is more evolutionary than we care to admit. For a long time working in the advertising industry, I was conscious that in every proposal we made we presented post-rationalisations as though they had been rational all along.

I am not suggesting that we try to solve problems completely at random, with no plan as to where we want to go, and nor do I mean

‡ In having various gods and goddesses of fortune – Ganesha, Tyche, Fortuna, etc. – ancient religions were perhaps more objective than modern rationalists, since they were not inclined to attribute all outcomes to individual human rational agency.

that data and rational judgement play no part in our deliberations. But in coming up with anything genuinely new, unconscious instinct, luck and simple random experimentation play a far greater part in the problem-solving process than we ever admit. I used to feel bad about presenting ideas as though they were the product of pure inductive logic, until I realised that, in reality, everything in life works this way. Business. Evolution by natural selection. Even science.

Even mathematicians, it seems, accept that the process of discovery is not the same as the process of justification. Cédric Villani is the holder of a Fields Medal, often described as the highest honour a mathematician can receive. He won his medal 'For his proofs of nonlinear Landau damping and convergence to equilibrium for the Boltzmann equation' and says, 'There are two key steps that a mathematician uses. He uses intuition to guess the right problem and the right solution and then logic to prove it.'

We have conflated the second part of this process with the first. We assume that the progress must appear as neat in the moment as it can be made to seem in retrospect, and we want ideas to be as straightforward in their formulation as in their analysis – instinct and luck can play no part in finding a solution. However, the experience of discovery simply does not bear out this approach. If it is true in physics and mathematics, it is probably even truer in questions of human behaviour.

In the foreword to a WPP annual report, Jeremy Bullmore uses as an example of discovery the legend of Archimedes in the bath – though he acknowledges that it may or may not be true, it illustrates an important truth nonetheless. So the story goes, King Hiero II, tyrant of Syracuse, provided his resident goldsmith with gold to make a votive crown for a temple, but when it arrived he suspected the goldsmith of having adulterated the gold with silver and of keeping the rest of it for his own purposes. He charged Archimedes with the task of establishing the truth – he knew the specific weight of gold, of course, but in order to determine whether the crown was made of pure gold, he needed to ascertain its volume.

A purely logical approach might have been to melt the crown down and form it into a brick – in which configuration its volume could readily be determined, but with the unfortunate consequence of destroying it. As Archimedes searched for a solution to the problem, Hiero's impatience began to grow. The question never entirely left Archimedes' mind, and it followed him into the bath, where he noticed that as he lowered his body into the bath, the water level rose, and as he began to leave the bath, the water fell. As Bullmore puts it, 'absolutely everything he observed or encountered was potentially relevant to that insistent problem', as if he knew that he had found a means of measuring the volume of complicated solids without knowing exactly how.

Bullmore notes that we tend to frown on those who admit their debt to intuition as opposed to carefully planned experiment. He imagines how Archimedes might have retrospectively described his discovery, if he were writing it up for a scientific journal:

'I approached this problem rationally. Since volume by definition implies space occupied, I reasoned that space occupied within a liquid allowed for the measurement of the volume of that liquid both before and after the immersion of a solid. It follows that the difference between the two, which I shall call "displacement", must precisely equal the volume of the solid immersed. Thereafter, the only requirement was the choice of a vessel of the requisite size and of a shape that was readily susceptible to conventional linear measurement.'

Bullmore's point, of course, is that this type of account is useful to validate an idea or discovery, but as an explanation of how the idea came about in the first place, it's quite false. He further contends that our tendency to attribute our successes to a planned and scientific approach and to play down the part of accidental and unplanned factors in our success is misleading and possibly even limits our scope for innovative work.

It is time to ask another stupid question: What is reason actually for? This may seem absurd, but in evolutionary terms it is far from trivial. After all, as far as we know, every other organism on the planet survives perfectly well without such a capacity. It is true that reason seems to have given us remarkable advantages over other animals – and it is unlikely that we could have produced many of our technological and cultural successes without it. But, in evolutionary terms, these must be a by-product, because evolution does not do long-term planning.§

Hence we must look for some other reason why we have such advantages, and we must also ask whether reason is designed to help us make most decisions or whether it evolved for some other purpose. It's true that we consciously believe our actions are guided by reason, but this does not mean that they are – it may simply be evolutionarily advantageous for us to believe this.

One astonishing possible explanation for the function of reason only emerged about ten years ago: the argumentative hypothesis¶ suggests reason arose in the human brain not to inform our actions and beliefs, but to explain and defend them to others. In other words, it is an adaptation necessitated by our being a highly social species. We may use reason to detect lying in others, to resolve disputes, to attempt to influence other people or to explain our actions in retrospect, but it seems not to play the decisive role in individual decision-making.

In my view, this theory has much to commend it. For one thing, it explains why individuals use reason so sparingly, selectively and above all self-servingly. It explains why we are good at contriving reasons for positions we already hold, or for decisions we have already made. And it explains confirmation bias, which leads people to seek out and absorb only that information which supports an existing belief. It also explains 'adaptive preference formation', where we change our perception of reality in order to depict

§ For instance, it does not think, hey, let's add a feature to the brain, so a million years later we can have an Apollo Program.

¶ First advanced by Dan Sperber and Hugo Mercier, and described in full in *The Enigma of Reason* (2017).

ourselves in a better light. In this model, reason is not as Descartes thought, the brain's science and research and development function – it is the brain's legal and PR department.

Understanding this theory seems important, first of all because it might help us see what human reason can and can't do well.** It might also help us understand how the misuse or overuse of reason can backfire. Collective, self-serving argument can work well when people are in possession of all the pertinent facts, which is why it works well in the physical sciences, when all the pertinent variables are known, and can be numerically expressed. However, in the social sciences this simply does not apply – it is impossible to quantify many of the important psychological factors which people care about, and there are no SI units for what really matters.

In the physical sciences, cause and effect map neatly; in behavioural sciences it is far more complex.

Cause, context, meaning, emotion, effect.

** For instance, we can be immune to sound arguments when they clash with an emotional predisposition, or when we don't like or trust the person making them.

1.18 THE OVERUSE OF REASON

One explanation for why apparently logical arguments may be ineffectual at changing people's minds, and why they should be treated with suspicion, is that it is simply too easy to generate them in the real world. As with 'GPS logic' it is possible to construct a plausible reason for any course of action, by cherry-picking the data you choose to include in your model and ignoring inconvenient facts. As I said earlier, the people who lost the Brexit referendum in the UK, and the Democrats who lost to Donald Trump in the US, both feel that their respective campaigns had the better arguments, but you would have to be a very committed Remainer or Democrat not to notice that the field in which they were prepared to argue in both cases was spectacularly narrow.

The more data you have, the easier it is to find support for some spurious, self-serving narrative. The profusion of data in future will not settle arguments: it will make them worse.

1.19 AN AUTOMATIC DOOR DOES NOT REPLACE A DOORMAN: WHY EFFICIENCY DOESN'T ALWAYS PAY

Business, technology and, to a great extent, government have spent the last several decades engaged in an unrelenting quest for measurable gains in efficiency. However, what they have never asked, is whether people like efficiency as much as economic theory believes they do. The 'doorman fallacy', as I call it, is what happens when your strategy becomes synonymous with cost-saving and efficiency; first you define a hotel doorman's role as 'opening the door', then you replace his role with an automatic door-opening mechanism.

The problem arises because opening the door is only the *notional* role of a doorman; his other, less definable sources of value lie in a multiplicity of other functions, in addition to door-opening: taxi-hailing, security, vagrant discouragement, customer recognition, as well as in signalling the status of the hotel. The doorman may actually increase what you can charge for a night's stay in your hotel.

When every function of a business is looked at from the same narrow economic standpoint, the same game is applied endlessly. Define something narrowly, automate or streamline it – or remove it entirely – then regard the savings as profit. Is this, too, explained by argumentative thinking, where we would rather win an argument than be right?

I rang a company's call centre the other day, and the experience was exemplary: helpful, knowledgeable and charming. The firm was a client of ours, so I asked them what they did to make their telephone operators so good. The response was unexpected: 'To be perfectly honest, we probably overpay them.'

The call centre was 20 miles from a large city; local staff, rather than travelling for an hour each day to find reasonably paid work, stayed for decades and became highly proficient. Training and recruitment costs were negligible, and customer satisfaction was astoundingly high. The staff weren't regarded as a 'cost' – they were a significant reason for the company's success.

However, modern capitalism dictates that it will only be a matter of time before some beady-eyed consultants pitch up at a board meeting with a PowerPoint presentation entitled 'Rightsizing Customer Service Costs Through Offshoring and Resource Management', or something similar. Within months, either the entire operation will be moved abroad, or the once-happy call centre staff will be forced on to zero-hour contracts. Soon nobody will phone to place orders because they won't be able to understand a word they are saying, but that won't matter when the company presents its quarterly earnings to analysts and one chart contains the bullet point: 'Labour cost reduction through call centre relocation/downsizing'.

Today, the principal activity of any publicly held company is rarely the creation of products to satisfy a market need. Management attention is instead largely directed towards the invention of plausible-sounding efficiency narratives to satisfy financial analysts, many of whom know nothing about the businesses they claim to analyse, beyond what they can read on a spreadsheet. There is no need to prove that your cost-saving works empirically, as long as it is consistent with standard economic theory. It is a simple principle of business that, however badly your decision turns out, you will never be fired for following economics, even though its predictive value lies somewhere between water divining and palmistry.

Take something called 'quad-play'. Economic orthodoxy these days demands that all mobile phone networks must also offer broadband, landlines and pay TV, that all those offering pay TV must likewise offer broadband, mobile telephony and landlines, and so on. The 'economic'* rationale for this is that, by offering all four together, you can enjoy back-office efficiency, economies of scale and price leadership; in economic models, it follows that whoever is the cheapest supplier of all four services will dominate the market. In the real world, however, quad-play is about as popular as a shit sandwich. The human brain has been calibrated by evolution not to pursue economic optimisation and risk systemic disaster. Quad-play places four eggs in one basket, which makes us feel vulnerable: refuse to pay that £250 data-roaming charge from your jaunt to Tenerife and one company can cut off your mobile, television, broadband and landline. And besides, the last thing anyone wants is an aggregated monthly reminder of what all the costs adds up to.†

Has business abandoned its traditional and socially useful role, where competing businesses tested divergent theories of how best to satisfy customer needs, with the market passing judgement on their efforts? It sometimes seems to have been reduced to a kind of monotheistic religion of efficiency where, provided you can recite the approved managerial mantras about economies of scale and cost savings to your financial overlords, no further questions will be asked.

Years ago, I had breakfast with the chief executive of one of Britain's largest companies, who arrived fresh from a grilling by City analysts. To those of you unfamiliar with modern business, the reason they were unhappy with him might seem strange; his company sold a product that was both the most expensive product on the market and also had the highest market share. What could possibly be wrong with that? You might have expected the analysts to thank him, but instead they claimed there was no way that the most expensive product in a category could also be the market

* i.e. stupid.

† I once tallied up what I spend each month on broadband, landline, mobile phones and pay TV, and my wife had to talk me down from a ledge.

leader and proposed that he must either drop the price of the product or accept that its market share was going to fall. I checked today, and seven years later, the product is still priced at a premium, and it has an even higher market share now than it did back then.

So much for economic orthodoxy – in fact, it is not uncommon for premium-priced products to have a high market share, as any of those financial analysts might have realised had they reached into their pockets to find an iPhone‡ or the key to an Audi. But to them it was more important that the company should act in a way that was consonant with economic theory than that it should succeed in supplying a large number of people with a superior product.

A year ago my own employer, with no consultation at all, moved every worldwide employee – over 70,000 people in total – to a new email platform, in the space of a single weekend. Many users felt it was palpably worse than the one that preceded it, but 'at least now it could be managed centrally'. What terrified me was that no tests were performed to see what effect the new platform might have on productivity. Our 70,000 people might each spend three or more hours every day involved with email, messaging or calendaring tasks, and so a platform that was only 5 per cent slower would result in a spectacular loss of productive time.

But no tests were performed, because the purpose of the activity, rather than to improve productivity, was to be able to tell a plausible story to analysts that we were making 'IT savings through back-office consolidation'. In the event, the platform has improved

‡ It was the same economic orthodoxy that prompted financial analysts to encourage Apple to launch the ill-fated iPhone 5c, a plasticky variant of the classic iPhone. The argument was that, without a low-cost model, Apple would fail to capture adequate market share. The product failed; anyone who couldn't afford a new iPhone had already solved the problem by buying or inheriting an older iPhone, not by using a manifestly inferior version.

since we adopted it, but the fact that a cost-saving decision could be made without any consideration of the hidden risks to efficiency was nonetheless alarming. Why are large commercial organisations adopting this ideological approach to business? That was supposed to be the weakness of communism.

It is a never-mentioned, slightly embarrassing but nevertheless essential facet of free market capitalism that it does not care about reasons – in fact it will often reward lucky idiots. You can be a certifiable lunatic with an IQ of 80, but if you stumble blindly on an underserved market niche at the right moment, you will be handsomely rewarded. Equally you can have all the MBAs money can buy and, if you launch your genius idea a year too late (or too early), you will fail.

To people who see intelligence as the highest virtue, this all seems hopelessly unmeritocratic, but that's what makes markets so brilliant: they are happy to reward and fund the necessary, regardless of the quality of reasoning. Perhaps people don't 'deserve' to be rewarded for being lucky, but a system that did not ensure the survival of lucky accidents would lose most of its value. Evolutionary progress, after all, is the product of lucky accidents. Similarly, a system of business that kept empty restaurants, say, open through subsidy, simply because there seemed to be some good reasons for their continued existence, would not lead to happy outcomes.

The theory is that free markets are principally about maximising efficiency, but in truth, free markets are not efficient at all. Admiring capitalism for its efficiency is like admiring Bob Dylan for his singing voice: it is to hold a healthy opinion for an entirely ridiculous reason. The market mechanism is loosely efficient, but the idea that efficiency is its main virtue is surely wrong, because competition is highly inefficient. Where I live, I can buy groceries from about eight different places; I'm sure it would be much more 'efficient' if Waitrose, M&S, Lidl and the rest were merged into one huge 'Great Grocery Hall of The People'.[§]

§ I am equally confident that the shop would be terrible.

The missing metric here is semi-random variation. Truly free markets trade efficiency for market-tested innovation that is heavily reliant on luck. The reason this inefficient process is necessary is because most of the achievements of consumer capitalism were never planned and are explicable only in retrospect, if at all. For instance, very few companies ever tested the effects of offshoring their call centre operations to countries with low labour costs – it simply became the fashionable thing to do, such was the level of enthusiasm for cost-reduction.

The following is a perfect illustration of the tendency of modern business to pretend that economics is true, even when it isn't. London's West End theatres often send out emails to people who have attended their productions in the past, to encourage them to book tickets, and it was the job of an acquaintance of mine who worked as a marketing executive for a theatre company to send out these emails. Over time, she learned something that defied conventional economic rules; it seemed that if you sent out an email promoting a play or musical, you sold *fewer* tickets if you included an offer for reduced-price tickets with the email. Conversely, offering tickets at the full price seemed to *increase* demand.

According to economic theory, this makes no sense at all, but in the real world it is perfectly plausible. After all, any theatre selling tickets at a discount clearly has plenty to spare, and from this it might be reasonable to infer that the entertainment on offer isn't all that good. No one wants to spend £100–£200 on tickets, a meal, car-parking and babysitting, only to find that you would have had more fun watching television at home; in avoiding discounted theatre tickets, people are not being silly – they are showing a high degree of second-order social intelligence.

Despite my friend's discovery, her colleagues continued to demand that she discount tickets. She patiently explained to them that any discount would reduce the demand, so that they would end up selling fewer tickets at a lower price, but they would insist that she included a discount anyway. They persisted in acting this way because, even though it was empirically the wrong thing to do, in economic terms it sounded logical. If 30 per cent of the seats failed to sell at a discounted price, it was assumed that they would not have sold at a higher price. If, by contrast, she hadn't offered a discount and 20 per cent of the seats had not sold, she could have been blamed. People's motivations are not always well-aligned with the interests of a business: the best decision to make is to pursue rational self-justification, not profit. No one was ever fired for pretending economics was true.

At the beginning of this book I looked at a charity appeal where the rational, logical improvements failed and the irrational, or psycho-logical ones succeeded. How many more solutions might we discover if we stop trying to solve everything to the satisfaction of our pre-frontal cortex, while ignoring the rest of our brains? That is what the next section hopes to uncover.

PART 2: AN ALCHEMIST'S TALE (OR WHY MAGIC REALLY STILL EXISTS)

2.1 THE GREAT UPSIDE OF ABANDONING LOGIC – YOU GET MAGIC

In the late Middle Ages, science took a wrong turn and came to the wrongheaded conclusion that alchemy didn't work. People had struggled for years to turn base metals into gold; when they found they couldn't do this in the way they expected, they gave up.

Later on, Newton really didn't help the cause by filling our heads with thermodynamics and the conservation of energy – where science hopelessly misled us was that it imbued in all of us the idea that you can't create something out of nothing. It taught us that you can't create a valuable metal out of a cheap one, or that you can't create energy in one place or form without destroying it somewhere else. While all this is perfectly true in the narrow sphere of physics, it is hopelessly wrong when it comes to the very different business of psychology.

In psychology these laws do not apply: one plus one can equal three.

Later on, economists got their own version of the same depressing idea that nothing can be created or destroyed. 'There's no such thing as a free lunch,' they said. The sad consequence is that no one believes in magic any more. Yet magic does still exist – it is found in the fields of psychology, biology and the science of perception, rather than in physics and chemistry. And it can be created.

The advertising agency J. Walter Thompson used to set a test for aspiring copywriters. One of the questions was simple: 'Here are two identical 25-cent coins. Sell me the one on the right.' One successful candidate understood the idea of alchemy. 'I'll take the right-hand coin and dip it in Marilyn Monroe's bag.* Then I'll sell you a genuine 25-cent coin as owned by Marilyn Monroe.'†

In maths it is a rule that 2 + 2 = 4. In psychology, 2 + 2 can equal more or less than 4. It's up to you.

We don't value things; we value their meaning. What they are is determined by the laws of physics, but what they mean is determined by the laws of psychology.

Companies which look for opportunities to make magic, like Apple or Disney, routinely feature in lists of the most valuable and profitable brands in the world; you might think economists would have noticed this by now.

Wine tastes better when poured from a heavier bottle. Painkillers are more effective when people believe they are expensive. Almost everything becomes more desirable when people believe it is in scarce supply, and possessions become more enjoyable when they have a famous brand name attached.

Sadly, no one in public life believes in magic, or trusts those who purvey it. If you propose any solution where the gain in perceived value outweighs the attendant expenditure in money, time, effort or resources, people either don't believe you, or worse, they think you are somehow cheating them. This is why marketing doesn't get any credit in business – when it generates magic, it is more socially acceptable to attribute the resulting success to logistics or cost-control.

* She was presumably alive at the time.

† A weirder alternative would be to devalue the coin on the left by lending it to Jeffrey Dahmer or Fred West. Most people wouldn't want it then – although it would probably find a buyer on eBay.

Ethically principled as it may sometimes seem, this aversion to magic brings huge problems; the ingrained reluctance even to entertain magical solutions results in a limitation in the number of ideas that people are allowed to consider. It is because of this that governments are usually confined to pulling on the twin levers of legal compulsion and economic incentive, while ignoring solutions which may be more cost-effective or less coercive. For instance, the UK Government's recent decision to spend £60 billion on a new high-speed railway to link London, Birmingham and Manchester is a case in point. The case for this expenditure is twofold: partly it's about saving time by means of new, faster trains, and partly it's about creating additional capacity.‡

However, the problem is cost. £60 billion is obviously a lot of money, and then there's the time it will take to build the new line. It's true that the new trains will knock around an hour off every journey, with a typical trip to Manchester reduced to something like 70 minutes from the current time of 2 hours and 10, but we'll have to wait till the end of the 2020s to enjoy this gain.§ Waiting ten years to save 60 minutes is hardly a compelling proposition. So I suggested a magical alternative that would reduce the journey time to Manchester by around 40 minutes *and* increase the capacity of the existing trains, all in the space of six months and at a trivial cost of around £250,000.

The trick I used was simple. Don't look at the logistics of the problem, look at it from the perspective of a passenger. To reduce journey times by 40 minutes, you don't have to reduce the amount of time people spend on the train – which is in any case the most enjoyable part of their journey – you could simply reduce the amount

‡ This may come as a surprise to American readers, but the use of trains in the UK has been growing for some years – indeed, more Britons travelled by train in recent years than at any time since the 1920s.

§ I'll have long retired by the time it comes into effect.

of time they waste *waiting for the train*. Provided their end-to-end journey is 40 minutes quicker, they've saved 40 minutes.

This would be easy to do. At the moment, most people buy an advance ticket to travel from London to Manchester or Birmingham, which gives a considerable saving on the cost of the journey, but requires you to travel on a specified train – and if you miss this train, the ticket is worthless. As a result, people typically allow a wide margin of error in fear of missing their train, and turn up at Euston station about 45 minutes before their designated train is due to depart. In those 45 minutes, two earlier trains will typically leave the station, and these generally have empty seats on them.

All you need to do to cut 40 minutes off the journey time, I explained, is to create a mobile app that would allow you to board one of the two earlier trains whenever spare seats are available, in return for a small voluntary payment. Obviously this wouldn't always work, as sometimes the earlier trains will be full, but most of the time it would be an easy way to allow people to cut 20–40 minutes off their time at the station. It would also have the additional benefit of increasing the capacity of the network, because otherwise empty seats would now be occupied, and the ones on the later train could be re-sold.

As far as I know, no one has taken this suggestion seriously – it does not fit into transport analysts' narrow, metric-driven conception of what improvement might look like. Their only conception of time-saving applies to time spent in motion – the means by which they aim to improve things are too narrowly defined.¶

¶ Similarly, the best way to improve air travel probably lies with faster airports, not faster aircraft.

2.2 TURNING LEAD INTO GOLD: VALUE IS IN THE MIND AND HEART OF THE VALUER

The reason the alchemists gave up in the Middle Ages was because they were looking at the problem the wrong way – they had set themselves the impossible task of trying to turn lead into gold, but had got it into their heads that the value of something lies solely in *what it is*. This was a false assumption, because you don't need to tinker with atomic structure to make lead as valuable as gold – all you need to do is to tinker with human psychology so that it feels as valuable as gold. At which point, who cares that it isn't *actually* gold?

If you think that's impossible, look at the paper money in your wallet or purse; the value is exclusively psychological. Value resides not in the thing itself, but in the minds of those who value it. You can therefore create (or destroy) value it in two ways – either by changing the thing or by changing minds about what it is.

One contention in this book is that nearly all really successful businesses, as much as they pretend to be popular for rational reasons, owe most of their success to having stumbled on a psychological magic trick, sometimes unwittingly. Google, Dyson, Uber, Red Bull, Diet Coke, McDonald's, Just Eat, Apple, Starbucks and Amazon have all deliberately or accidentally happened on a form of mental alchemy. Alongside these great successes, we should also remember a group of companies you have never heard of: those that failed. Often their business ideas were perfectly logical, but they failed because they didn't contain any alchemy.

Preoccupied as they were with the hopeless idea of 'transmutation' – the transformation of one element into another – the alchemists failed to experiment with the *rebranding* of lead. Perhaps they could have added a mystery ingredient or polishing technique to make it slightly shinier and named the result 'Black Gold'. Or, better still, they might have used the French trick of creating artificial scarcity through topography and provenance* and called their special lead something like 'Or de Sable de Lyon'. This regional monopoly would have maintained rarity and made their product more expensive than boring old gold. All that would have been needed then was a major PR event: perhaps a king somewhere could have been induced to commission a crown made of it, by which point the job would have been half done. Some craftsmen at the time had indeed managed to do exactly this through a mixture of skill, scarcity and branding; for instance Limoges enamelware, which contained mostly copper, was at the time, pound for pound, considerably more valuable than gold.†

---

* Notable examples include champagne and burgundy.

† Modern printer ink, ounce for ounce, is more expensive than gold.

2.3 TURNING IRON AND POTATOES INTO GOLD: LESSONS FROM PRUSSIA

In nineteenth-century Prussia, a glorious feat of alchemy saved the public exchequer, when the kingdom's royal family managed to make iron jewellery more desirable than gold jewellery. To fund the war effort against France, Princess Marianne appealed in 1813 to all wealthy and aristocratic women there to swap their gold ornaments for base metal, to fund the war effort. In return they were given iron replicas of the gold items of jewellery they had donated, stamped with the words 'Gold gab ich für Eisen', 'I gave gold for iron'. At social events thereafter, wearing and displaying the iron replica jewellery and ornaments became a far better indication of status than wearing gold itself. Gold jewellery merely proved that your family was rich, while iron jewellery proved that your family was not only rich but also generous and patriotic. As one contemporary observed, 'Iron jewellery became the fashion of all patriot women, thus showing their contribution in support of the wars of liberation.'

Yes, precious metals have a value, but so does meaning, the addition of which is generally less expensive and less environmentally damaging. After all, thinking what is gold jewellery is actually *for* reveals it to be an extremely wasteful way of signalling status. But it was perfectly possible, with the right psychological ingredients, to allow iron to do this job just as well. Psychology 1, Chemistry 0.

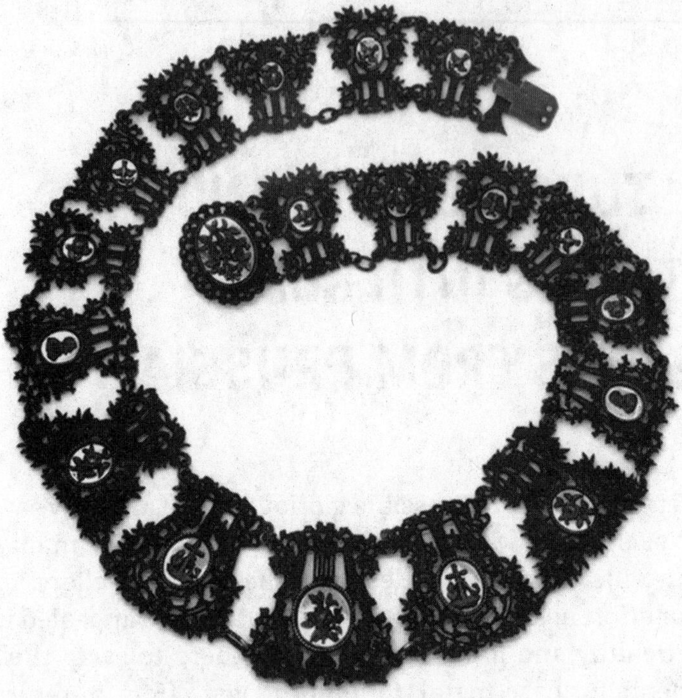

'Gold gab ich für Eisen'. How alchemy made iron jewellery a higher-status ornament than gold.

One eighteenth-century monarch, Frederick the Great, used the same magic in the promotion of the potato as a domestic crop, transforming something worthless and unwanted into something valuable through the elixir of psychology. The reason he wanted eighteenth-century Prussian peasants to cultivate and eat the potato was because he hoped that they would be less at risk of famine when bread was in short supply if they had an alternative source of carbohydrate; it would also make food prices less volatile. The problem was that the peasants weren't keen on potatoes; even when Frederick tried coercion and the threat of fines, they simply showed no interest in eating them. Some people objected because the potato was not mentioned in the Bible, while others argued that, since dogs wouldn't eat potatoes, why should humans?

So, having given up on compulsion, Frederick tried subtle persuasion. He established a royal potato patch in the grounds of his palace, and declared that it was to be a royal vegetable, that could only be consumed by members of the royal household or with royal permission.* If you declare something highly exclusive and out of reach, it makes us all want it much more – call it 'the elixir of scarcity'. Frederick knew this and so posted guards around his potato patch to protect his crop, but gave them secret instructions not to guard the patch too closely. Curious Prussians found they could sneak into the royal potato patch and could steal, eat and even cultivate this fabulously exclusive vegetable for themselves. Today, the potato – which is unsurpassed as a source of nutrients and energy – is as popular in Germany as it is everywhere else.†

---

* Rather like Cheddar cheese in seventeenth-century England, or swan meat in modern Britain.

† I have heard rumours that various hip-hop clothing brands have deployed a similar strategy to Frederick; by making it relatively easy for their clothes to be shoplifted, the stolen clothes ended up being worn by people who were significantly cooler than the people who paid full price. Similarly, some breweries certainly design beer glasses in the hope that they are pilfered. As one client told me, 'They are getting a free glass costing, perhaps 30p; we're getting a free advertisement in their kitchen.'

Potatoes on the grave of Frederick the Great.

2.4 THE MODERN-DAY ALCHEMY OF SEMANTICS

But surely this kind of alchemy no longer works today? Well, have you ever eaten Chilean sea bass?* It is the product of a particular sort of alchemy, 'The Alchemy of Semantics'. The $20 slice of fish that graces plates in high-end restaurants under the name 'Chilean sea bass' actually comes from a fish that for many years was known as the Patagonian toothfish. No one is going to pay $20 for a plate of Patagonian toothfish – call it Chilean sea bass, however, and the rules change. An American fish wholesaler called Lee Lentz had the idea, even though, strictly speaking, most of the catch doesn't come from Chile and the toothfish isn't even related to the bass.†

* People who really hate fish should consider skipping the next few pages.

† The importance of naming also extends to universities. A May/June 1999 article, 'Overrated & Underrated' by John Steele Gordon, in *American Heritage* magazine rated Elihu Yale the 'most overrated philanthropist' in American history, arguing that the college that became Yale University was successful largely because of the generosity of a man named Jeremiah Dummer. For some reason, the trustees of the school did not want it known by the name 'Dummer College'.

Dishonest as it may seem, Lentz's action in fact sits within a long tradition of rebranding seafood. Monkfish was originally called goosefish, orange roughy was once called slimehead, and sea urchins were once whore's eggs. More recently, a similar thing happened to pilchards. Caught off the Cornish coast before being salted and shipped all over Europe, they had been a delicacy for centuries, until the advent of domestic refrigeration and freezing caused the appetite for salted fish – at least outside of Portugal – fall away. 'The market was dying fast as the little shops that sold them closed down,' says Nick Howell of the Pilchard Works fish suppliers in Newlyn. 'I realised I needed to do something about it.' Fortunately, Nick thought creatively. He discovered that what the Cornish often called the pilchard was related to the fish that was served, with lemon and olive oil, to British tourists in the Mediterranean as a fashionable sardine.‡ So he changed the name from the pilchard, a name redolent of ration food,§ to the 'Cornish sardine'. Next, a supermarket buyer who called to ask for French sardines was deftly switched to buying 'pilchards from Cornwall'. A few years ago Nick successfully petitioned the EU to award Cornish sardines Protected Designation of Origin (PDO) status, and the result was extraordinary: the *Daily Telegraph* reported in 2012 that sales of fresh sardines at Tesco had rocketed by 180 per cent in the past year, an increase that was partly explained by a huge

‡ The Latin name is *sardina pilchard.*

§ And, perhaps even worse, of cat food. Pilchards were a common cat-food flavour in Britain: it does little to endow a food with the allure of scarcity when it is given to pets. Interestingly, many foods which have the allure of luxury today were treated as more or less disposable at times and in places where they were abundant: domestic staff in Scotland in the nineteenth century were known to demand that it was written into their contract that 'they not be fed salmon more than three times a week'.

increase in the sales of 'Cornish sardines'. This rebranding exercise had reinvigorated the entire Cornish fishing industry.

Cornish sardines are another example of geographical alchemy at work.¶ Merely adding a geographical or topographical adjective to food – whether on a menu in a restaurant or on packaging in a supermarket – allows you to charge more for it and means you will sell more. According to research from the University of Illinois at Urbana-Champaign, descriptive menu labels raised sales by 27 per cent in restaurants, compared to food items without descriptors.

On menus, there seems to be more money in adjectives than in nouns. Even adjectives that have no precise definition such as 'succulent' can raise the popularity of items. The Oxford experimental psychologist Charles Spence has published a paper on the effect the name of a dish has on diners. 'Give it an ethnic label such as an Italian name,' he says, 'and people will rate the food as more authentic.'** We make far more positive comments about a dish's appeal and taste when it is garlanded with an evocative description: 'A label directs a person's attention towards a feature in a dish, and hence helps bring out certain flavours and textures.'

Never forget this: the nature of our attention affects the nature of our experience.

Advertising†† often also works in this way. A great deal of the effectiveness of advertising derives from its power to direct attention to favourable aspects of an experience, in order to change the experience for the better. Strangely, there is one form of enhancement to a menu that seems to be the kiss of death: adding photographs of

¶ For international readers, Cornwall is a rural and beautiful county in the far south-west of England, with strong foodie associations.

** A gelateria can charge more than an ice-cream parlour.

†† The word comes from the Latin '*anima advertere*', or 'to direct attention'.

dishes to a menu seems to heavily limit what you can charge for them. Opinion is divided on why. Some people think that the practice is strongly associated with downmarket restaurants, while others believe that attractive photographs may raise expectations too high, leading to inevitable disappointment when the real food arrives. It is certainly interesting to me that many cult burger restaurants, including Five Guys and In-N-Out, have simple textual menus and no photographs, while McDonald's uses photographs on its LCD screens extensively. Does this limit their power to charge a premium?‡‡

‡‡ Japan is an exception to this rule. In Japan, not only do upmarket restaurants display high-quality photographs of the food, but there is a skilled area of craftsmanship where highly paid people make models of sushi and other foods to display in restaurant windows. I would not recommend this approach to a restaurant in London, Paris or New York.

2.5 BENIGN BULLSHIT – AND HACKING THE UNCONSCIOUS

It is easy to disparage alchemy as bullshit. And to be frank, some of what I say will probably later be shown to be bullshit. But much of it – the renaming of fish, the lavish addition of geographical provenance to menu items, the rebranding of iron – can be placed under the category of 'benign bullshit', because the same technique that works for fish can also solve much more significant problems. For instance, how can we encourage more women to pursue tech careers? Or, to change the question, how can we prevent tech careers from seeming unappealing to women? One college has found the answer. In 2006, Maria Klawe, a computer scientist and mathematician, was appointed president of Harvey Mudd College in California. At the time, only 10 per cent of the college's computer science majors were women. The department devised a plan, aimed at luring in female students and making sure they actually enjoyed their computer science initiation, in the hopes of converting them to majors.

A course previously entitled 'Introduction to programming in Java' was renamed 'Creative approaches to problem solving in science and engineering using Python'.* The professors further divided the

* Words like 'creative' and 'problem solving' just sounded less nerdy.

class into groups – Gold for those with no coding experience and Black, for those with some coding experience.† They also implemented *Operation Eliminate the Macho Effect*, in which males who showed off in class were taken aside and told to desist. Almost overnight, Harvey Mudd's introductory computer science course went from being the most despised required course to the absolute favourite.

That was just the beginning. Improving the introductory course obviously helped, but it was also important to ensure that women signed up for another class. The female professors took the students to the annual Grace Hopper Conference, an annual 'celebration of women in technology'. It was an important step in demonstrating that there was nothing weird or anti-social about women working in tech. Finally, the college offered a summer of research for female students to apply their new-found talents to something useful and socially beneficial. 'We had students working on things like educational games and a version of Dance Dance Revolution for the elderly. They could use computer technology to actually work on something that mattered,' says Klawe.

As is often the case with nudges, they had a multiplicative effect and the movement snowballed. After the first four-year experiment, the college had quadrupled its female computer science majors in a short space of time, from ten per cent to 40 per cent. Notice that there were no quotas involved – everything was voluntary, and no one found their freedom to choose impaired. It is simply good marketing applied to a problem.

The invention of the 'designated driver' was an even cleverer use of semantics and naming to create a social good. The phrase, meaning the person who is nominated to stay sober in order to drive his friends home safely, was a deliberate coinage that spread with the active support of Hollywood who agreed to use it in selected episodes of popular sitcoms and dramas. The phrase first originated in Scandinavia, was adopted by the Hiram Walker distillery in

† Note this crafty use of colours.

Canada to promote the responsible use of its alcohol products and was then deliberately brought into the US, at the bidding of the Harvard Alcohol Project.

Once you can casually ask, 'Who's going to be the DD on Friday?' it's easy to see how this behaviour becomes much easier to adopt, and it's also much easier for the sober person to defend their sobriety when anyone offers them a drink. In Belgium and the Netherlands, he (or she) simply explains I can't drink tonight, I'm Bob' – a Dutch acronym‡ for *Bewust Onbeschonken Bestuurder* or 'deliberately sober driver'. In both cases, creating a name for a behaviour implicitly creates a norm for it.

It is interesting to consider how many more benign behaviours might be made possible through semantic invention. I have always thought, for instance, that the word 'downsizing', which is used not only as a euphemism for redundancies, but in another sense refers to the voluntary decision by 'empty nesters' to move to a smaller and more manageable home, is a very useful coinage. It allows older people in needlessly large homes to portray their move to a smaller house as a choice born out of preference, rather than – as it may otherwise be assumed to be – a compromise born of financial necessity. Create a name, and you've created a norm.§

---

‡ Technically a 'backronym', since the name Bob preceded the longer form version.

§ In Britain, much-despised student loans would be perceived very differently if they were simply reframed as 'the graduate tax'.

2.6 HOW COLOMBIANS RE-IMAGINED LIONFISH (WITH A LITTLE HELP FROM OGILVY AND THE CHURCH)

When Hurricane Andrew hit the south-eastern US in 1992, it was the worst hurricane in US history. It caused incalculable damage both to property and to the environment; however, its biggest environmental effect, perhaps, was not the loss of a species, but the opposite. In South Florida, the hurricane burst a large coastal aquarium tank, releasing an unwelcome species of fish into the Gulf of Mexico and the Caribbean.

The lionfish comes from the tropical waters around Indonesia. Though beautiful to look at, it is a voracious predator of other fish, and is able to eat as many as 30 in half an hour. Furthermore, one female lionfish can produce over two million eggs per year, which was a particular problem in the Caribbean, where it has no natural predators. The decimation of local species threatened the environment and the economics of Colombia, much of which depends on fishing. It was also destroying the ecology of coral reefs. This was when some colleagues of mine borrowed an idea from Frederick the Great; Ogilvy & Mather in Bogotá decided that the solution was to create a predator for the lionfish – humans. The simplest and most cost-effective way to rid Colombia's waters of lionfish was to encourage people to eat them, which would encourage anglers to catch them. The agency recruited the top chefs in Colombia and encouraged them to create lionfish recipes for the best restaurants.

As they explained, a lionfish is poisonous on the outside but delicious on the inside, so they created an advertising campaign titled 'Terribly Delicious'. Working with the Colombian Ministry of the Environment, they generated a cultural shift by turning the invader into an everyday food. Lionfish soon appeared in supermarkets. Some 84 per cent of Colombians are Roman Catholic, so they asked the Catholic Church to recommend lionfish to their congregations on Fridays and during Lent. That additional element – recruiting the Catholic Church – was the true piece of alchemy. Today, indigenous fish species are recovering and the lionfish population is in decline.*

* Will this action alone be enough to kill off the lionfish? Probably not. However, it isn't necessary to eradicate the lionfish entirely to solve the problem – you simply need to keep its numbers below a certain threshold. A study released by Oregon State University found that in reefs where numbers were kept below 'threshold density' native fish populations increased by between 50 and 70 per cent, while in regions where there had been no effort to fight the invasive species, local fish continued to disappear.

2.7 THE ALCHEMY OF DESIGN

We know how to design physical objects to fit the shape of the human hand quite well. Unless you are a small child, or you are staying in a pretentious boutique hotel where everything is chosen to signal 'Hey, we're totally different,'* door handles are generally found at a height and in a shape that suits your frame. Good designers know to create objects which work well with our evolved physique, even if those parts of our bodies originally evolved for entirely different purposes; we did not evolve hands to hold car steering wheels, nor did we develop sticky-out ears in order to stop our spectacles falling off, but good designers know that such features can be useful for purposes other than those for which they were selected.

* One New York hotel (The W, in Times Square, I think) has a sign pointing to the elevators that uses the British term 'Lifts' as a fancy point of difference. Of course, if you are British yourself this doesn't really work – I suppose it is equivalent to British property advertisements now using 'apartment' rather than 'flat', in signalling cosmopolitanism.

In general, the physical world *is* designed fairly well. There are some scandalous exceptions,[†] but for the most part we do a reasonable job, because we accept that our bodies are a funny shape and design objects to work with them. Even better, in wealthier countries we now design the environment to suit people who are less fit than the average person, or do not have full use of all their limbs. This practice has been driven by groups campaigning on behalf of disabled people, and in a few cases probably has been slightly overengineered, though it has also brought many unexpected benefits to people other than those for whom it was primarily intended – for, in reality, all of us are disabled some of the time. If you are carrying heavy luggage, staircases are almost unusable. If you are carrying a cup of coffee, you have effectively lost the use of a hand. If you wear glasses but do not have them on, you are visually impaired.

Even when you are designing for the able-bodied, it is a good principle to assume that the user is operating under constraints. This is why a door handle is better than a door knob: it allows you to open a door with your elbow – either because you do not have any hands, or because your hands are holding cups of tea. The provision of wheelchair ramps at airports may benefit the owners of rolling suitcases almost as much as wheelchair users. Subtitles

† Why is marble considered a sensible flooring material for hotel bathrooms? I suspect the answer here is again, 'because of what it signals': marble is a scarce material, and thus expensive; it therefore conveys the idea that the hotel has spared no expense. It is a pity, however, that this display comes at the expense of your safety. If hoteliers must install needlessly posh bathrooms, they should at least use a costly *non-slip* material. No one seems to compile much data on accidents in hotel bathroom: I know, however, of four colleagues who have been hospitalised from such falls, some of them even while sober.

meant for the 'hard of hearing' are likewise useful if you want to watch television in a bar or airport, or while your children are asleep.‡

This sort of design can be sound business practice: some years ago, British Telecom introduced a telephone for the visually impaired, with enormous buttons. To their great surprise, this model became their bestselling product; it was a phone that able-bodied people were easily able to use while lying on their side in bed and not wearing their glasses. OXO Good Grips is a highly successful manufacturer of kitchen utensils that applies this principle to the wider world: Sam Farber started the company because his wife suffered from arthritis and had difficulty using kitchen implements – very rapidly, the ease of use and comfort of its well-designed products expanded its popularity into the able-bodied population. It's worth remembering that products designed for people with imperfect grip also work well for people with wet hands – common in cooking.§

Eventually most physical objects, by a form of natural selection, acquire a shape and function matched to our evolved preferences and instincts. After a few decades, this principle came to extend to software interface design.¶ Finger gestures such as pointing, clicking, pinch-to-zoom and so forth have become the default modes of interaction with technological devices, simply because they closely

‡ Or when Hollywood awards Oscars to mumbling actors.

§ With compliments to Adam Morgan, author of *Eating the Big Fish* and *Beautiful Constraint*.

¶ The doyen of this school of thinking is Don Norman, author of the apparently brilliant book *The Design of Everyday Things*. I say apparently, because my paperback edition is printed in a ludicrously small type, and I cannot read more than a few pages at a time. Not Don's choice, I suspect.

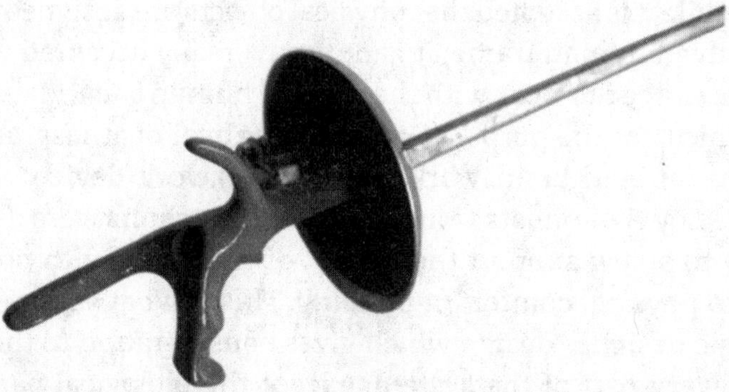

This fencing sword was designed for fencers who had lost fingers, though the handle design was eventually adopted by professional fencers with the full number of fingers.

resemble the instinctive movements we have been making for a few hundred thousand years or more.**

The three principal modern forms of media consumption device, the laptop/desktop computer, the tablet and the mobile phone, are also products of the human form. Of the millions of hypothetical possibilities†† there are, effectively, three comfortable modes for the human body: 1) standing up, 2) lying down and 3) sitting upright. The three devices with which we access digital content mirror these fairly well. A mobile device for when you are moving about, a tablet for laid-back use and a laptop or desktop for when you are sitting up at desk.‡‡

** The declarative pointing gesture is uniquely human, although domestic dogs seem to have evolved an innate understanding of human finger-pointing. From an early age, they will look in the direction of the extended human finger: a few hundred thousand years before Steve Jobs, dogs had evolved their own version of a graphical user-interface – point-and-whistle, rather than point-and-click.

†† For instance, hanging upside down from your toes.

‡‡ Starbucks owes a large part of its revenues to hiring out horizontal surfaces (tables) to laptop users under the guise of selling coffee.

But while it is accepted that physical objects are designed around the evolved human frame, it is not universally accepted that the world is shaped to work with the evolved human *brain*. Mainstream economics, for the purposes of mathematical neatness, assumes that the human brain works like a clockwork device. A world designed by economists would be one where chairs were designed merely to stably support the weight of the sitter, with no regard given to physical comfort or padding. This is what you might call 'aspergic design' – design which gives consideration to the working of every part of the system, except the biological part.[§§] But our brains have also evolved, and also shift in shape, just like our bodies.

A knowledge of the human physique is considered essential in designing a chair, but a knowledge of human psychology is rarely considered useful, never mind a requirement, when someone is asked to design a pension scheme, a portable music player or a railway. Who is the Herman Miller of pensions, or the Steve Jobs of tax-return design? These people are starting to emerge – but it has been a painfully long wait. If there is a mystery at the heart of this book, it is why psychology has been so peculiarly uninfluential in business and in policy-making when, whether done well or badly, it makes a spectacular difference.

§§ An extreme example of this is often found in car-park design, where up and down ramps are placed at 90 degrees to the direction of travel of your car, to minimise the amount of concrete needed, even though this requires people repeatedly to execute a difficult manoeuvre with a high chance of damage to their vehicle. By contrast, if you want to see the work of the Steve Jobs of car park design, visit Bloomsbury Square in London, where the underground car park is a double-helix shape; it is possible to go all the way to the bottom and back up again with your steering wheel in one position.

2.8 PSYCHO-LOGICAL DESIGN: WHY LESS IS SOMETIMES MORE

Economic logic suggests that more is better. Psycho-logic often believes that less is more. Akio Morita came from a Japanese family that had been involved in the production and sale of soy and miso sauce since the mid-seventeenth century. With his business partner Masaru Ibuka he founded Sony (as the Tokyo Telecommunications Engineering Company) in 1946. Magnetic tape recorders were the company's first area of focus, followed by the first fully transistorised pocket radio.* But his greatest moment of genius was perhaps the creation of the Sony Walkman, the ancestor of the iPod.

To anyone born after 1975 there is nothing outlandish about people walking around or sitting on a train wearing headphones, but in the late 1970s this was a very odd behaviour indeed; comparable to the use of an early cellphone in the late 1980s, when to use one in public carried a high risk of ridicule.† In market research, the

---

* Strictly speaking these radios were not pocket-sized, but in an early manifestation of his genius, Morita ordered shirts with outsized pockets for his employees. If you can't make the radio smaller, make the pocket larger.

† I can also remember seeing my first jogger in the 1970s – I assumed for a moment that he was being pursued by some unseen assailant.

Walkman aroused very little interest and quite a lot of hostility. 'Why would I want to walk about with music playing in my head?' was a typical response, but Morita ignored it. The request for the Walkman had initially come from the 70-year-old Ibuka, who wanted a small device to allow him to listen to full-length operas on flights between Tokyo and the US.‡

When the engineers came back, they were especially proud. Not only had they succeeded in achieving what Morita had briefed them to create – a miniature stereo cassette player – they had also managed to include a recording function. I imagine they were crestfallen when Morita told them to remove that extra function. The technology involved,§ given the economics of mass production, would have added no more than a few pounds to the final purchase price, so why would you *not* add this significant extra?¶ Any 'rational' person would have advised Morita to go with the engineers' advice, but according to multiple accounts, Morita vetoed the recording button.

This defies all conventional economic logic, but it does not defy psycho-logic. Morita thought the presence of a recording function would confuse people about what the new device was for. Was it for dictation? Should I record my vinyl record collection onto cassette? Or should I record live music? In the same way that McDonald's omitted cutlery from its restaurants to make it obvious how you were supposed to eat its hamburgers, by removing the recording function from Walkmans, Sony produced a product that had a lower range of functionality, but a far greater potential to a change behaviour. By reducing the possible applications of the device to a single use, it clarified what the device was *for*. The technical design term for this

‡ Or perhaps it was Morita's own idea – accounts differ.

§ The innards of the Walkman were in part derived from the Sony Pressman, a miniature dictation machine much used by journalists.

¶ The first Walkman did in any case contain a microphone – but this was to allow a companion to talk to you while you were wearing headphones.

is an 'affordance', a word that deserves to be more widely known. As Don Norman observes:

'The term *affordance* refers to the perceived and actual properties of the thing, primarily those fundamental properties that determine just how the thing could possibly be used. [...] Affordances provide strong clues to the operations of things. Plates are for pushing. Knobs are for turning. Slots are for inserting things into. Balls are for throwing or bouncing. When affordances are taken advantage of, the user knows what to do just by looking: no picture, label, or instruction needed.'

Once you understand this concept, you can perhaps understand why Morita was right.** It is always possible to add functionality to something, but while this makes the new thing more versatile, it also reduces the clarity of its affordance, making it less pleasurable to use and quite possibly more difficult to justify buying.

The world is full of invisible intelligence of this kind. One defence I always make of traditional architecture is that it is user-friendly. A few years ago I was one of a group of people speaking at a conference in a brutalist 1960s building on the South Bank in London. All of us were wandering around the outside of the building, trying glass doors, unable to work out where we were supposed to go in. Say what you like about the British Museum, but nobody in 150 years has ever approached its classical portico and thought, 'Hmm, I wonder where the door is?'

Imagine for a second a door with a handle *and* a 'push plate', above which appears the word PULSH. Had Sony produced a

** Although credit for the idea for the Sony Walkman (and a $10m payment) was eventually ceded by Sony to Andreas Pavel, who had earlier patented a 'stereobelt' while living in Brazil, I am fairly confident that Morita or Ibuka should be credited with the important idea of removing recording functionality from the device.

Any idea where the door is?

Walkman with a recording function, this is what it would have been. A PULSH – a thing whose function was not unambiguously clear. The Walkman also exploits a clear psychological heuristic, or rule of thumb – 'the jack-of-all-trades-heuristic', whereby we naturally assume that something that only does one thing is better than something that claims to do many things. Similarly, when we hear 'sofa-bed', we instinctively think of an item of furniture that is not great as a sofa and not much good as a bed, either. And some of you may have encountered a spork, an unsatisfactory spoon, which is rather less use as a fork.

Those of a scientific inclination will – quite fairly – make the point that there is no evidence that removing the recording function from the Walkman was a good idea. There is no parallel universe in which a multi-function model was launched and subsequently flopped. It is also true that recording functionality was added to later versions of the Walkman, though this occurred after the function of the device had been widely adopted and understood.[††] However, all I can rely on here for evidence is a recurrent pattern of events – it is surprisingly common for significant innovations to emerge

†† In the same way, the first iPhone was largely comprehensible because people were already familiar with the iPod.

from the removal of features rather than the addition. Google is, to put it bluntly, Yahoo *without* all the extraneous crap cluttering up the search page, while Yahoo was, in its day, AOL *without* in-built Internet access. In each case, the more successful competitor achieved their dominance by removing something the competitor offered rather than adding to it.

Similarly, Twitter's entire *raison d'être* came from the arbitrary limitation on the number of characters it allowed. Uber originally did not allow you to pre-book cars. Highly successful publications such as the *Week* effectively take the world's newspapers and make them digestible by removing a lot of extraneous content; McDonald's deleted 99 per cent of items from the traditional American diner repertoire; Starbucks placed little emphasis on food for the first decade of its existence and concentrated on coffee; low-cost airlines competed on the basis of what in-flight comforts you *didn't* get. If you want to offer ease of use – and ease of purchase – it is often a good idea not to offer people a Swiss Army knife, something that claims to do lots of things.[‡‡] With the notable exception of the mobile phone, we generally find it easier to buy things that serve a single purpose.

However, the engineering mentality – as at Sony – runs counter to this; the idea of removing functionality seems completely illogical, and it is extremely hard to make the case for over-riding conventional logic in any business or government setting, unless you are the chairman, chief executive or minister in charge. Although you may think that people instinctively want to make the best possible decision, there is a stronger force that animates business decision-making: the desire not to get blamed or fired. The best insurance against blame is to use conventional logic in every decision.

'No one ever got fired for buying IBM' was never the company's official slogan – but when it gained currency among corporate

‡‡ You may well own a Swiss Army knife but, if you do, I would guess you have only used it when nothing else is to hand.

buyers of IT systems, it became what several commentators have called 'the most valuable marketing mantra in existence'. The strongest marketing approach in a business-to-business context comes not from explaining that your product is good, but from sowing fear, uncertainty and doubt (now commonly abbreviated as FUD) around the available alternatives. The desire to make good decisions and the urge not to get fired or blamed may at first seem to be similar motivations, but they are, in fact, never quite the same thing, and may sometimes be diametrically different.

PART 3:
SIGNALLING

3.1 PRINCE ALBERT AND BLACK CABS

I mentioned earlier in this book that there are five main reasons why human behaviour often departs from what we think of as conventional rationality. The first of these is signalling, the need to send reliable indications of commitment and intent, which can inspire confidence and trust. Cooperation is impossible unless a mechanism is in place to prevent deception and cheating; some degree of efficiency often needs to be sacrificed in order to convey trustworthiness or to build a reputation.

For instance, in London I can put my two daughters into a car driven by a complete stranger and rely on them to be driven safely to their destination, because the stranger is driving a black cab. Before anyone can drive a black cab, he or she is forced to undergo a gruelling four-year initiation programme known as the Knowledge, for which they are required to memorise every street, major building and commercial premises within six and a half miles of Charing Cross Station, an area that includes 25,000 streets and 20,000 landmarks. This requires that they spend most of their spare evenings and weekends riding around on mopeds on test routes, before appearing in front of regular examination panels to test their knowledge of the quickest or shortest route between any two points; so gruelling is the process that it seems to enlarge the hippocampus of those who take part in it. In cabbie folklore, the model for the

Knowledge was first suggested by Prince Albert.* The test is certainly Teutonically stringent: over 70 per cent of the applicants either fail or drop out.†

As useful as it once was, many people feel the Knowledge has been made superfluous by the arrival of satnavs and Google Maps. Conventional economic thinking, obsessed with 'market efficiency', would argue that the Knowledge seems a 'barrier to entry' erected to maintain the scarcity of cab drivers. I was tempted to agree – but that was before I realised that the Knowledge had as much value as a signal than a navigation skill.

A market like the London taxi industry, where you almost never interact with the same person more than once, needs a high level of trust in order to work, and one way to establish this confidence is to demand serious proof of commitment before you are admitted to the trade. It is in the interests of all honest cab drivers to maintain a standard of trust; if only 0.5 per cent of cab journeys resulted in a rip-off or a mugging, faith in the whole system would evaporate and the entire business would collapse.

Medieval guilds existed for this reason. Trust is always more difficult to gain in cities because of the anonymity they afford, and guilds help to offset this problem. If it is costly and time-consuming to join one, the only people who enter are those with a serious commitment to a craft. Guilds are also self-policing; the upfront cost of being admitted adds to the fear of being ejected.

Commitment devices of this kind exploit the fact that, once you have made a commitment – whether in time, money or effort – there

* Whether this story is true or not, there is something slightly German about the notion that every tradesman should have a qualification.

† Americans should remember that there is no numbered grid system in London – every street is individually named and, to add further confusion, the same street name may reappear in different parts of the city.

is no way of reversing it. To put it another way, would I really trust a black cab driver with my daughters' safety if they could obtain a black cab licence merely through attending three or four evening classes and shelling out for a second-hand TomTom?

Reciprocation, reputation and pre-commitment signalling are the three big mechanisms that underpin trust. You can repeatedly use a small local firm that needs your loyalty, you can use a larger company with a brand reputation, such as Addison Lee or Green Tomato Cars, or you can trust someone who has made a big investment in getting a badge and stands to lose everything if he is caught cheating. If you don't believe this, go to Athens, where foreign taxi passengers are on average taken on a 10 per cent longer route than Athenian passengers. Try Seville, where I was menaced to pay an imaginary €20 '*suplemento aeropuerto*'. Or Rome, where a colleague of mine was mugged by his taxi driver.‡

Uber is a taxi company with different mechanisms to promote trust. There is a digital record of every journey, a rating system and, increasingly stringent checks on driver background. I'm not arguing that the Knowledge is the only solution to this problem, but what I am saying is that it is only partially of navigational value; a large part of its value is as a signalling device. It also means that black cab drivers tend to be highly experienced, because there would be little point in undergoing a four-year initiation if you were only planning to drive taxis as a stop-gap. In that respect, the Knowledge is an upfront expense that is proof of long-term commitment.

‡ The London system is not perfect: one taxi driver, John Worboys, was convicted in 2009 for 12 rapes and sexual assaults – and was suspected of committing many more, but such cases are very rare indeed. In London in 2016, 31 licensed taxi drivers were charged with sexual offences, and not a single black cab driver was among them.

3.2 A FEW NOTES ON GAME THEORY

Many things which do not make sense in a logical context suddenly make perfect sense if you consider what they *mean* rather than what they are. For instance, an engagement ring serves no practical purpose as an object. However, the object – and its expense – make it highly redolent with meaning; an expensive ring is a costly bet by a man in his belief that he believes – and intends – his marriage to last.

Now, you might expect a book of this kind to have a chapter about the Ultimatum Game* and other experimental, game-theoretic investigations into the nature of trust and reciprocation. This book contains no such chapter. The reason for that is that the Ultimatum Game is stupid, and so is the Prisoner's Dilemma: these games exist in a context-free, theoretical universe with no real-life parallels. They both posit the idea of the one-shot exchange, a transaction involving two strangers with no knowledge of the other's identity. In the real world, such transactions never take largely place – we

* The Ultimatum Game and Prisoner's Dilemma are both theoretical exercises that investigate how cooperation can be made to work. Google them, by all means, but remember they are artificial.

choose to buy things in shops, not from random strangers in the street.†

When we engage in transactions, we are generally aware of the other party's identity and can see clues to their commitment. For example, if I enter a shop in a town I have never previously visited, there is a slim chance that the shopkeeper may take my money and refuse to hand over the goods – he might be an imposter. However, let's now assume that the shop is called H. Jenkins & Sons and above the door is a notice that reads 'Established 1958'. The shopkeeper has clearly invested in his premises and stock, and it's unlikely he will have survived in business for several decades if his business model depended on ripping off the local population.‡ He is at a point where the loss of his reputation will be far more costly than whatever he might gain by refusing to honour our exchange, which allows me to derive trust from the context in which the exchange takes place. Upfront investment is proof of long-term commitment, which is a guarantor of honest behaviour. Reputation is a form of skin in the game: it takes far longer to acquire a reputation than to lose one.

If you wish to make an Ultimatum Game work so that everybody would cooperate, allow me to propose a simple mechanism: you

† There is only one instance where anyone I've met engaged in a one-shot, high-value exchange with someone whose identity they did not know and, predictably, it was a disaster. A friend had moved from England to Australia and wished to buy a second-hand car. The man selling the car asked to meet him in a supermarket car park. As my friend explained, 'This seemed weird, but I was newly arrived in Australia and just assumed that that was the way things were done here.' He foolishly completed the transaction, and the car turned out to be stolen.

‡ If the shop mainly serves tourists, all bets are off – at least until the advent of TripAdvisor, you could safely con tourists with impunity.

simply demand that, before anyone is allowed to play, they are required to memorise 25,000 streets and 20,000 landmarks in London. This will take four years of their life. At that point, all you need is a simple mechanism to ensure that cheaters can be expelled from the game. Under such circumstances no one will wish to cheat, since it would place them at risk of being thrown out of the game before they have recouped the effort they invested in gaining entry in the first place.§

There remains one problem, which is the possibility that people may cheat immediately before they leave the game.¶ In theory, there is nothing to stop a London taxi driver from taking an appallingly roundabout route when carrying the last passenger of his working life. If he loses his badge then, he sacrifices no future revenue, since there is no future revenue to lose. Whether intentionally or not, London taxi drivers seem to have a solution to this. A few years ago I took a £15 ride in central London and at the end of the journey proffered a £20 note to the young driver. 'Can't take it, mate, you're my first ever ride,' he said. 'What? Seriously? Why not?' 'It's a tradition – you don't take money for your first ever fare.' I loved this at the time, but it now occurs to me that this tradition is an unbelievably elegant one – if, in the course of my life in London, I also end up being picked up once by a cab driver as his last ever fare and he rips me off a bit, me and the London cabbies will effectively be quits.

---

§ This may be why companies are so eager to hire graduates. Someone who has invested money and time in the quest for a good job is unlikely to void their entire investment by wandering around your office, stealing laptops.

¶ I think quite a few people *do* nick things on their last day at work – although they often refer to these items as 'souvenirs'. Nicking things on your first day at work is a much riskier enterprise.

3.3 CONTINUITY PROBABILITY SIGNALLING: ANOTHER NAME FOR TRUST

I mentioned that before the prospect of repeat custom is something that keeps businesses honest, but there is another conclusion we can draw – that you can signal that you are an honest business by showing that your business model relies on repeat custom.

What do the following things have in common?

1. The fact that large, carnivorous fish desist from eating useful fish such as wrasse, which clean the parasites from their bodies
2. The posh rope-handled carrier bags you get when you spend an appreciable amount of money on clothes or cosmetics
3. The free extra scoop of fries they give you at Five Guys
4. The fortune spent on a wedding
5. Your modest minibar charges that are waived by a hotel
6. The marble and oak lavishly used in bank branches
7. That fancy training course your company sent you on
8. A lavish advertising campaign
9. A free glass of limoncello given to you by a restaurant after your meal
10. Investment in a brand

Looked at from the perspective of simple, short-term economic rationality, none of these behaviours makes sense. A bank could conduct its business perfectly well from a Portakabin. Those rope-handled bags are expensive but they aren't even waterproof. Limoncello isn't cheap, and a lot of people don't like it. Was the training course really worth £5,000?

All these things only make sense if we assume that some signalling is going on – they are examples of a behaviour which is costly in the short term and which will only pay off, if at all, in the long term. They are thus – if nothing else – reliable signals that the person, animal or business engaging in that behaviour is acting on the basis of long-term self-interest rather than short-term expediency.

This distinction matters a great deal. Unlike short-term expediency, long-term self-interest, as the evolutionary biologist Robert Trivers has shown, often leads to behaviours that are indistinguishable from mutually beneficial cooperation. The reason the large fish does not eat its cleaner fish is not because of altruism but because *over the long-term*, the cleaner fish is more valuable to it alive than dead. The cleaner fish in turn could cheat by ignoring the ectoparasites and eating bits of the host fish's gills instead, but its long-term future is better if the big fish becomes a repeat customer.* What keeps the relationship honest and mutually beneficial is nothing other than the prospect of repetition.

In game theory, this prospect of repetition is known as 'continuation probability', and the American political scientist Robert Axelrod has poetically referred to it as 'The Shadow of the Future'. It is agreed by both game theorists and evolutionary biologists that the prospects for cooperation are far greater when there is a high expectation of repetition than in single-shot transactions. Clay Shirky has even described social capital as 'the shadow of the future at a societal scale'. We acquire it as a means of signalling our commitment to long-term, mutually beneficial behaviours, yet some businesses

* And they do. Fish, it seems, exhibit surprising brand loyalty towards individual cleaner fish.

barely consider this at all – procurement, by setting shorter and shorter contract periods, may be unwittingly working to reduce cooperation.

Yet there are, when you think about it, two contrasting approaches to business. There is the 'tourist restaurant' approach, where you try to make as much money from people in a single visit. And then there is the 'local pub' approach, where you may make less money from people on each visit, but where you will profit more over time by encouraging them to come back. The second type of business is much more likely to generate trust than the first.

How might we distinguish the second type of business from the first? Well, the scoop of extra fries you get at Five Guys is one such gesture – an immediate (if modest) expense with a deferred pay off, and a reliable signifier that the business is investing in a repeat relationship and not milking a single transaction. Likewise, when your company pays your salary each month, it says you are worth that money for now; when it sends you on an expensive training course, it signals that it is committed to you for at least a few years.†

If fish (and even some symbiotic plants) have evolved to spot this sort of distinction, it seems perfectly plausible that humans instinctively can do the same, and prefer to do business with brands with whom they have longer-term relationships. This theory, if true, also explains some counterintuitive findings in customer behaviour: it has long surprised observers that, if a customer has a problem and a brand resolves it in a satisfactory manner, the customer becomes a more loyal customer than if the fault had not occurred in the first place. Odd, until you realise that solving a problem for a customer at your own expense is a good way of signalling your commitment to a future relationship. The theory of 'continuation probability' would also predict that, when a business focuses narrowly on

† It is widely known in the training community that the biggest gain from a company's investment in training comes in the form of staff loyalty.

short-term profit maximisation, it will appear less trustworthy to its customers, something that seems all too plausible.

Remember that hanging over every human interaction is an unspoken question: 'I know what I want from this transaction. But what's your interest in this exchange? And can I trust you to fulfil your promises?' We do not need to know that the other party is honest; we simply need to know that they will behave in the transaction as an honest person would. In a closer community, you may simply be able to establish a reputation for all-round honesty, and that's that. No bank manager could risk cheating a single customer in the 1950s, because if one customer found out they'd been cheated, the bank manager's reputation was destroyed through the entire town.

There are many forms of expenditure – of money and of effort – which make sense within the context of a relationship, but which make no sense in a single transaction. Small acts of discretionary generosity, such as waiving a charge when a passenger has bought slightly the wrong kind of train ticket or a complimentary chocolate at the end of a meal are regarded by customers as reassuring indicators of trustworthiness; we correspondingly see the absence of such signals as being a cause for concern.

One of the reasons why customer service is such a strong indicator of how we judge a company is because we are aware that it costs money and time to provide. A company which is willing to spend time after you have bought and paid for a product to make sure you are not disappointed with it is more likely to be trustworthy and decent than one which loses all interest in you as soon as the cheque has cleared. The same applies in interpersonal relations; being rude isn't so different from being polite, but it requires less effort. Politeness demands that we perform hundreds of little rituals, from opening doors to standing up when someone enters the room, all of which are more effortful than the alternative. By such oblique means we convey that we care about their opinion – and about our reputation.

3.4 WHY SIGNALLING HAS TO BE COSTLY

Twenty years ago, a colleague and I were working on a small-but-important advertising brief. Our task was to send a letter to a few thousand senior IT professionals, asking them to try out the Microsoft Windows NT 32-bit server software in advance of its wider release. We could have simply sent them all a letter by second-class post telling them what the product was and what we were offering them – that would have conveyed information, but no meaning. Instead we produced an elaborate box containing a variety of bits and pieces including a free mouse-mat* and a pen, inside gratuitously expensive packaging.

We did that to convey to them not only that the product existed, but that it was a highly significant launch, to which Microsoft had committed a great deal of money. We also needed to convey the fact that very few people were in the privileged position of being able to test the software for free. We could have said that in the letter, but it would have meant nothing. It's what's called 'cheap talk', something that anybody selling anything can tell you – it is merely a claim, not an item of proof. Indeed to send an 'exclusive invitation' by second-class post (even worse, by bulk mail) would

* This *was* the 1990s.

have been self-contradictory – 'We're sending this exclusive invitation to loads of people.' (This is why no truly exclusive club can ever advertise in the mass media.)

So the mail pack we produced was elaborate, and of a kind that would have been uneconomical to produce in bulk – and it also won an award. But I most recall it because we worked on it with a Midwestern account director called Steve Barton, who said something telling when he briefed the project. 'Look', he said, 'I'd like you to produce a stand-out piece of creative work here. But if you can't, what I'd like you to do is write them a really nice one-page letter – and we'll send it out by FedEx.' Steve was effectively describing what biologists call 'costly signalling theory', the fact that the meaning and significance attached to a something is in direct proportion to the expense with which it is communicated.

Imagine you are the recipient – could you throw away a FedEx envelope unopened? I think it is safe to say that you could not. What we wanted from the recipient was not simply the ingestion of information: it was attention, conviction and a sense of import, something that an economically rational 50p stamp could never obtain but that a £10 FedEx envelope could. In the end, the campaign was hugely successful. Almost everyone opened the package and read the contents – and over 10 per cent tried the product, which required a significant amount of effort. In 2018 a digital rationalist would suggest that the way to reach several hundred senior IT people would be via Facebook, or by email – two options which were mercifully not available to us in the mid 1990s. He would be rationally right but emotionally dead wrong.†

† On one occasion, an advertiser used the principle of costly signalling to run a television advertisement with a target audience of only two or three hundred people. The people in question were chief executives of the British subsidiaries of large American multinationals. Almost all of them at the time were Americans, so in

Bits deliver information, but costliness carries meaning. We do not invite people to our weddings by sending out an email. We put the information (all of which would fit on an email – or even a text message) on a gilt embossed card, which costs a fortune. Imagine you receive two wedding invitations on the same day, one of which comes in an expensive envelope with gilt edges and embossing, and the other (which contains exactly the same information) in an email. Be honest – you're probably going to go to the first wedding, aren't you?‡

the late 1980s this company ran an advertisement promoting itself during the British transmission of the Super Bowl on Channel 4, at a time when American football was almost unknown in the UK, which meant that airtime was extraordinarily cheap. For these Americans, of course, it would be the one programme on UK TV they were sure to watch in the course of the year. For the Americans, it was 'a Super Bowl ad'. For us Brits it was just some welcome ad-break filler in the midst of an obscure and incomprehensible sport.

‡ Sorry, but you are. With the second invitation, there's the nervous suspicion that there might be a cash bar – I mean if they can't pay for a stamp, they're unlikely to splash out on an exciting range of boutique gins, are they?

3.5 EFFICIENCY, LOGIC AND MEANING: PICK ANY TWO

'*Credo quia absurdum est*', said Saint Augustine, supposedly – 'I believe it because it is ridiculous.' He was talking about Christianity, but it is equally true of many other facets of life: we attach meaning to things precisely because they deviate from what seems sensible. It is hardly surprising that we have evolved to invest more significance in unusual, surprising or unexpected stimuli and signals than to routine, everyday 'noise'. As a result, like any social species, we need to engage in ostensibly 'nonsensical' behaviour if we wish to reliably convey meaning to other members of our species.

The psychophysicist Mark Changizi has a simple evolutionary explanation for why water 'doesn't taste of anything': he thinks that the human taste mechanism has been calibrated not to notice the taste of water, so it is optimally attuned to the taste of anything that might be polluting it. If water tasted like Dr Pepper, it would be easier for sensory overload to drown out the hint of 'dead sheep', which would alert us to the fact that a carcass was decaying in a pool five hundred yards upstream. Water 'tastes of nothing', so we notice the smallest thing which deviates from this. You can try a similar experiment with young children. Feed them their favourite food, but add a subtle herb or spice. They will find it revolting, because the slight deviation from what they expect alarms them into believing it is somehow unsafe.*

My contention is that our perception is calibrated more widely in this way. We notice and attach significance and meaning to those things that deviate from narrow, economic common sense, precisely *because* they deviate from it. The result of this is that the pursuit of narrow economic rationalism will produce a world rich in goods, but deficient in meaning. In architecture this has produced modernism, a style with a marked absence of decoration or 'spurious' detail, and a corresponding loss of 'meaning'.† My secret hope is that, with the 3D-printing of buildings becoming possible, a certain Gaudí-ness may reappear in twenty-first-century buildings.‡

---

* Young children develop conservative food tastes around the age at which they learn to crawl, which prevents them from experimenting riskily with their diet.

† Nicholas Gruen, an economist friend of mine, recently visited Barcelona and on seeing Gaudí's Sagrada Familia remarked, 'God, if it hadn't been for modernism, the whole of the twentieth century could have looked like this!'

‡ Modernism isn't particularly efficient as an architectural style, by the way. Arches are better than beams for supporting a load, and flat roofs are awful in engineering terms. But modernist architecture, like economics and management consultancy, is good at creating the appearance of efficiency.

3.6 CREATIVITY AS COSTLY SIGNALLING

If you can't afford to pour money into the paper or the printing of your wedding invitation, you can use another scarce commodity that I'll call 'creativity', although it encompasses a variety of talents: design, artistry, craftsmanship, beauty, photographic talent, humour, musicality or even mischievous bravery. A handmade birthday card can be cheaper and yet still more moving than an expensive bought one – but it has to involve a level of effort.*
A video recording of a self-composed song as a wedding invitation could, with enough talent and tolerable production values, be sent by email, but a straightforward and factual unfunny email invitation just isn't the same – it has no creativity and is just a statement of fact.

The meaning in these things derives from the consumption of some costly resource – which, if not money, may be talent, or effort, or time or skill or humour or, in the case of risqué humour, bravery.†
But it has to contain something costly, otherwise it is just noise.

---

* After the age of four, you can't just scribble on a piece of paper.

† *Fountain* (1917) by Marcel Duchamp qualifies as art perhaps through bravery.

THE FAVOUR OF A REPLY IS KINDLY REQUESTED
BY MAY FIFTH

________ ENTHUSIASTICALLY ATTEND
________ REGRETFULLY DECLINE
________ REGRETFULLY ATTEND
________ ENTHUSIASTICALLY DECLINE

Bravery and wit can be a form of costly signalling.

Effective communication will always require some degree of irrationality in its creation because if it's perfectly rational it becomes, like water, entirely lacking in flavour. This explains why working with an advertising agency can be frustrating: it is difficult to produce good advertising, but good advertising is only good *because* it is difficult to produce. The potency and meaningfulness of communication is in direct proportion to the costliness of its creation – the amount of pain, effort, talent (or failing that, expensive celebrities or pricey TV airtime) consumed in its creation and distribution. This may be inefficient – but it's what makes it work.

Quite simply, all powerful messages must contain an element of absurdity, illogicality, costliness, disproportion, inefficiency, scarcity, difficulty or extravagance – because rational behaviour and talk, for all their strengths, convey no meaning. When Nike chose to use Colin Kaepernick, the American football player who had instigated the practice of kneeling rather than standing for the US national anthem before games, as the figurehead for their 2018 campaign, it was an example of a kind of costliness through bravery. He was not an expensive choice – his career was in limbo – but he was a brave one, since he was so closely identified with the NFL protests against police brutality. As this campaign demonstrated, meaning is conveyed by the things we do that are not in our own short-term self-interest – by the costs that we incur and the risks we take.

One of the most important ideas in this book is that it is only by deviating from a narrow, short-term self-interest that we can generate anything more than cheap talk. It is therefore impossible to generate trust, affection, respect, reputation, status, loyalty, generosity or sexual opportunity by simply pursuing the dictates of rational economic theory. If rationality were valuable in evolutionary terms, accountants would be sexy. Male strippers dress as firemen, not accountants; bravery is sexy, but rationality isn't. Can this theory be extended further? For instance, is poetry more moving than prose because it is more difficult to write?[‡] And is music more emotionally potent than normal speech because it is more difficult to sing than to talk?[§]

‡ Sorry, that should be 'Prose is easier to write than verse; Its persuasive powers are therefore worse.' Although poetry is assumed to be in decline, I was delighted to read recently that Wayne Rooney writes love poems to his wife, Coleen. It's hard to convey devotion in an email.

§ Sorry that should be, 'When you add a tune to something, there is one great consequence: we will somehow give it meaning, even if it makes no sense,' sung to the tune of Beethoven's 'Ode to Joy'.

3.7 ADVERTISING DOES NOT ALWAYS LOOK LIKE ADVERTISING: THE CHAIRS ON THE PAVEMENT

A few years ago, a coffee shop opened on a fairly busy road a mile or so from my house. There were about twenty seats inside, and a few benches on the pavement outside. It wasn't a bad coffee shop, but in time it failed. Some new people took over, following what seemed to be an identical formula, but they failed too.

The third owners to take over the premises therefore seemed overconfident in trying the same formula, yet they miraculously created a successful business. The food and the prices did not seem to be any different from that of their predecessors. In fact the only thing they changed seemed trivial: they bought more attractive chairs and tables, and placed them outside at the start of the day, as well as a waist-level gauze fence which surrounded the chairs, making a kind of terrace. This was less efficient than the old benches, since this moveable (and therefore thievable) furniture had to be stored away at the end of each day, and replaced every morning.

However, I think it was precisely this change that was the reason for the new shop's success. I mentioned that the café was on a busy road – in fact, to anyone concentrating on their driving, its existence wouldn't have been immediately obvious. Even if you did spot the sign saying 'coffee', it was far from clear, when no one was sitting outside, whether the coffee shop was open – you could have spent five minutes finding a parking space only to find it was closed.* The old benches that permanently sat outside were meaningless as an indicator of whether the shop was open. By contrast, the new chairs and the fence, which might have been stolen or have blown away if left unattended, were a guarantee that the shop was open – no one who had closed their shop would have gone home and left them in the street.

'Oh, come on,' I hear you say. 'This is all very well in theory, but nobody driving down an A-road consciously calculates the probability that a café is open by assessing the portability of the furniture outside.' In one sense you are right, but they don't do it consciously – they do it instinctively. And to make such calculations we use mental processes, which take place beyond the reach of conscious awareness. We draw unconscious inferences from environmental cues everywhere we go, without having the slightest awareness that we are doing so – it is thinking without thinking that we are thinking.

These mental processes are psycho-logical rather than conventionally logical, and rely on a different set of rules to those we adopt when we use conscious reasoning, but they are not necessarily irrational, given the conditions under which our brains have evolved. Our brains did not evolve to make perfect decisions using mathematical precision – there wasn't much call for this kind of thing on the African savannah. Instead we have developed the ability to arrive at pretty good, non-catastrophic decisions based on limited, non-numerical information, some of which may be deceptive. Far from being irrational, the inferences we are able to draw just from seeing chairs outside a café are surprisingly clever, once you uncover the reasoning behind them.

A sign that said 'Open' could be a meaningless claim, because someone could simply have forgotten to turn the sign to 'Closed' –

* Anyone familiar with provincial British tea and coffee shops will know that they follow the most eccentric opening hours in the known universe.

and in any case it would be hard to read from a car. A neon sign that said 'Open' would be a more reliable indicator, since someone leaving the shop would probably switch it off to save electricity.† But light, stackable chairs behind a windbreak – now that's a signal you can trust. In other words, the chairs act as an effective advertisement; the cost of their purchase and the daily effort entailed in arraying them outside the business and restacking them at the end of the day is a reliable signal of the existence of a functioning coffee shop, and one that is tacitly understood rather than consciously processed by human reason. Having worked in advertising for over 25 years, usually for large companies with big budgets, it still fascinates me how great an effect unconscious signalling can have on the fortunes of a tiny business. And more than that, it frightens me to think how many perfectly worthwhile businesses have failed that might not have done if they'd implemented a few trivial signals.‡

Relatively small businesses that might not be able to afford to advertise in any conventional sense, could transform their fortunes by paying a little attention to the workings of psycho-logic. The trick involves simply understanding the wider behavioural system within which they operate. Cafés could boost sales by improving their menu design. Many small shops are inadequately lit, and so passers-by assume they are closed – how much business do they lose as a result?§ Pubs are often needlessly intimidating because their windows are made of frosted glass, preventing people from looking inside before entering. Pizza delivery firms could differentiate themselves in a crowded market by agreeing to deliver tea, coffee, milk and toilet paper

† Though a neon sign would be better suited to an American diner than a British coffee shop.

‡ I know of one branch of John Lewis that could double their sales simply by placing a sign at the entrance to their car park.

§ The dying words of J. Sainsbury, founder of the Sainsbury's chain: 'Make sure the stores are kept well-lit'.

alongside a pizza. Restaurants might increase sales by allowing the kerbside collection of take-away meals – or by adding a sign which says 'parking at rear'.¶

In the unlikely event that either of the failing cafés had decided to appoint a management consultancy to solve their business woes, I doubt anyone would have suggested changing the furniture; doubtless they would have received a long list of recommendations covering all the left-brain facets of the business – pricing, stock control, staffing levels and so forth. Anything that could be included on a spreadsheet would be analysed, quantified and optimised, in order to increase efficiency. But no one would have mentioned the chairs.**

I will now take my idea one step further. Not only would we reliably infer from the presence of tables and chairs that the café is open, I also believe we go deeper still – I think we subliminally deduce that any place that goes to the trouble of erecting chairs on the street will serve coffee that, at the very least, is unlikely to be terrible. That seems a silly use of mental energy – surely the way to determine whether the coffee is good is to buy one and find out?

'I knew the coffee was going to be good because of the chairs,' sounds like a very silly sentence, but hold on a moment – maybe, using psycho-logic and a bit of social intelligence, we can identify a connection. For a start, someone who invests in new chairs and goes to the trouble of placing them on the pavement every day is not lazy, and has also invested in their business. Furthermore, they seem to expect their business to be a success – had they not, they

¶ My use of one local restaurant doubled when I discovered an obscure public car park hidden behind it.

** I have never worked for McKinsey, Bain or the Boston Consulting Group, so I may be doing them a great disservice, but I think I am safe in saying that you don't earn much kudos within those technocratic organisations by talking about furniture.

would not have undertaken the expense. The chairs don't promise perfection, but they are a reliable indicator of at least reasonable quality. The business owner who buys the windbreak and the chairs has probably also invested in a decent Gaggia machine, proper milk and coffee beans – and in training his staff. It suggests the owner, rather than playing the short game of immediate profit maximisation, is playing the long game, building a reputation and a loyal customer base – which will mean a cappuccino that is palatable at the very least.

Of course, you might have to be careful not to overdo this kind of signalling. Putting expensive armchairs outside might lead people – not unreasonably – to conclude that the establishment is also expensive. This question is a significant dilemma in supermarket design: the main factor which influences human price perception in shops is not, bizarrely, the actual prices charged, but the degree of opulence with which the store is fitted out.

If this emphasis on advertising seems excessive and self-serving, I sympathise – in fact, I thought this myself. However, it all depends how you define advertising; in nature, it is often necessary for something to present a persuasive message, and in a way that can't be faked. Information is free, but sincerity is not, and it isn't only humans who attach significance to messages in proportion to the costliness of their creation and transmission; bees also do it.[††]

†† As Cole Porter once said.

3.8 BEES DO IT

When signalling their enthusiasm for a potential nesting site, bees waggle about in an exponential relationship to its quality; the amount of energy they expend in the signalling of a potential nest site is proportional to their enthusiasm for it. But they also make use of expensive 'advertising', in order to decide where to devote their time and attention.

> **The advertisements which bees find useful are flowers – and if you think about it, a flower is simply a weed with an advertising budget.**

Flowers spend a great deal of their resources convincing customers that they are worth visiting. Their target audience is bees, or other insects, birds or animals that may help to pollinate the flower – a process that dates back at least to the time of the dinosaurs.* For the pollination process to be effective, the flower needs to convince the customers of its worth. To borrow the language of the *Michelin Guide*,

* The relationship between flowers and bees is technically called mutualism. I apologise if I mention bees rather a lot in this book, but mutualism is particularly revealing of the mechanisms by which honest cooperation can be instigated and sustained.

a flower can be 'vaut l'étape', 'vaut le détour' or 'vaut le voyage'; 'worth stopping at', 'worth going out of your way for' or 'a destination in itself'. To do this, the flower places a costly bet, offering a generous source of nectar that rewards bees for their visit, and encourages them to stay at the flower for long enough to collect pollen on their bodies for dispersal elsewhere. But this nectar is kept out of sight – how can the flower, at a distance, convince the bee of the existence of a reward which it cannot verify until it has already exerted time and effort?†

The answer is that they use 'advertising and branding' – they produce distinctive, hard-to-copy scents and large, brightly coloured petals. These are noticeable, but producing them is risky, as they may attract the attention of herbivores that might eat them. The distinctive scent and petals act as a reliable (though not infallible) proxy for the presence of nectar, which a bee can use to help decide whether the visit is worth it or not.

A plant which has sufficient resources to produce petals and scent is clearly healthy enough to produce nectar, but using its resources for distinctive display will only really pay off if bees visit more than once, or if they encourage other bees to join them – there is no point in advertising heavily up front if you only make one sale. When you come here, the display says, I'm betting that you're going to come back, or all my effort will have been wasted.

The system of information-sharing between the two species is also reliable – there is often a correlation between the size of petals and the supply of nectar. This saves a lot of wasted visits, because it means that a bee can tell from some distance away whether a plant is 'a destination in itself'. It also requires that the plant use their resources on being distinctive as well as noticeable. If any type of flower is a better source of nectar, this generosity will only be rewarded with 'customer loyalty' if bees can learn to recognise it and so choose to make repeat visits. If all flowers looked and smelled alike, any incentive they offered to the bee – more nectar, perhaps – would

† After all, I'm not driving 50 miles to a restaurant if I can't be fairly sure that the food is spectacular.

be ineffective, because the bee might not be able to distinguish between that kind of flower and other less rewarding plants. It is only by having a recognisable identity that a flower is able to improve the value exchange‡ and increase the chance of repeat visits.

I have used marketing jargon here, because what flowers need to establish in the minds of bees is, effectively, a brand. Why don't flowers cheat, by devising an alluring advertisement of huge petals, and then delivering no costly nectar? Well, sometimes they do – false advertising is common in orchids, which often seem to be the scam artists of the plant kingdom. At least one orchid species mimics the appearance (and smell) of female insect genitalia; many mimic food sources and some mimic other plants. But this can only work on a small scale§ – play that trick too often and insects will just learn to avoid you.

To put it another way, if there is the possibility of bees either refusing to return to a plant, or encouraging a wider boycott among their fellow bees, the resources spent on advertising through scent and colouration is an unrewarded cost. However, orchids are the tourist restaurants of the floral world – they rely on people visiting only once so are less worried about ripping off visitors, because they know they are never going to come back anyway. However, if there is a prospect of repeat visits or of positive reputational gains spreading among your prospective clientele,¶ it pays not to cheat. This mechanism isn't perfect: as with humans, it only works well

‡ By noticeable, I am not just referring to vision. Scent may be more important – and it also seems more difficult for other plants to mimic. But apparently 'Bees don't just recognise flowers by their colour and scent; they can also pick up on their minute electric fields,' a mechanism that was only recently discovered.

§ Which may be why such orchids are rare and tend to flourish only at the beginning of the season, before bees wise up.

¶ Just as TripAdvisor and other ratings mechanisms have changed the game here.

where there is the prospect of a regular repetition of exchange or a mechanism for reputational sharing. In categories where we only buy infrequently[**] and where we don't talk to each other about our satisfaction, it will break down.

Economists tend to dislike the idea of branding and are inclined to see it as an inefficiency, but then they might view a flower as an inefficient form of weed. The reason they might not understand the flower's extravagance in squandering its resources on producing scent and colour is that they don't fully understand what it is trying to do or the decision-making and information-transfer context in which it is trying to do it.

It is not any more irrational for human consumers to pay a premium for heavily advertised products than it is for bees to prefer to visit heavily 'advertised' flowers. It seems unlikely that a company would spend scarce resources advertising a product they believed to be bad – it would simply lead to the unpopularity of a bad product spreading more quickly. Moreover, a company with a long-established reputation for high-quality products has much more to lose from customer disappointment than a company with no reputation. To quote a Caribbean proverb, 'Trust grows at the speed of a coconut tree and falls at the speed of a coconut.' As with bees, this mechanism works because we are able to punish cheats: either an individual can never revisit, or a group can collectively boycott the brand through negative word of mouth (or in bees, though word of waggle dance).[††]

In advertising, a large budget does not prove a product is good, but it does establish that the advertiser is confident enough in the future popularity of the product to spend some of his resources promoting it. Since at the moment you make a purchase decision, the advertiser knows more about his product than you do, a costly

** Or once, in the case of a pension or a funeral plan, for instance.

†† Bees signal nectar and pollen sites to each other by an elaborate dance, where the direction of the dance signals the direction of a site worth visiting.

demonstration of faith by the seller may well be the most reliable indicator of whether something is at least worthy of consideration (remember the Knowledge with London taxi drivers). It also proves that the seller is in sufficient financial health to advertise in the first place. However, for this to work, you need stable distinctive identities, as well as laws to prevent manufacturers pretending their goods are those of someone else (this is called 'passing off' in commerce; in biology it is known as 'Batesian mimicry').‡‡

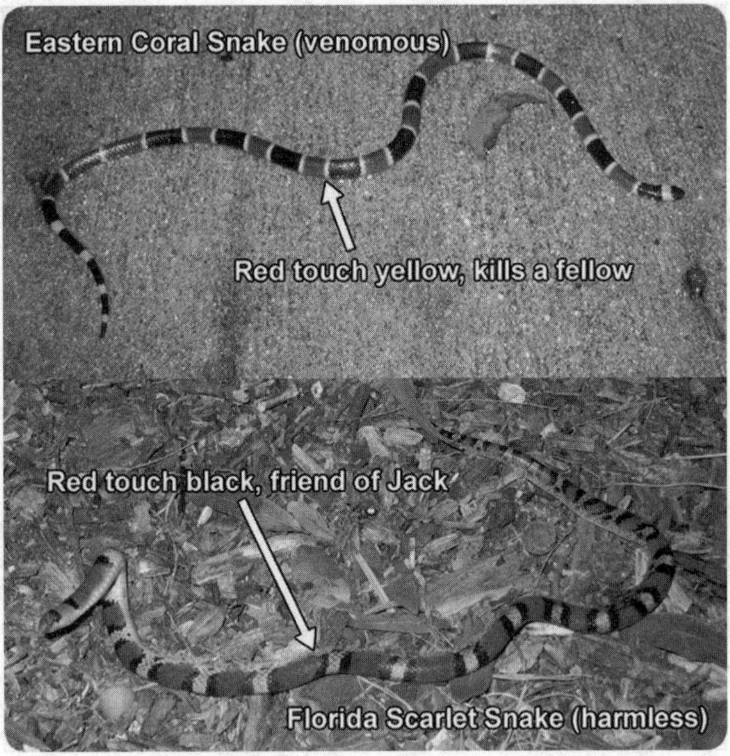
Eastern Coral Snake (venomous)
Red touch yellow, kills a fellow
Red touch black, friend of Jack
Florida Scarlet Snake (harmless)

Batesian mimicry in action.

‡‡ The kingsnake, which is harmless, mimics the coral snake, which is deadly. The rhyme to remember the difference is 'Red touching black, safe for Jack. Red touching yellow, kill a fellow.'

3.9 COSTLY SIGNALLING AND SEXUAL SELECTION

Costly signalling theory, which was first proposed by the evolutionary biologist Amotz Zahavi, is, I believe, one of the most important theories in the social sciences.* The idea of signalling and its role in sexual selection is necessary to explain many evolutionary outcomes, but it didn't always seem that way, not even to Charles Darwin. In a letter to a friend, Darwin remarked that 'the sight of a feather in a peacock's tail' made him 'physically sick'. The reason for this strange aversion was that the male peacock seemed a living refutation of the theory of evolution through natural selection – the idea that something so beautiful and yet so ostensibly pointless sat more easily with the idea of a divine creator than with the idea of natural selection. After all, a decorative tail in no way enhances fitness or survival and rather makes the peacock conspicuous to predators, and is an encumbrance when it needs to evade them. The ability to lurk unobtrusively in the shadows is an advantage both to predators and prey, but being highly visible *seems* to be a disadvantage to both.

It is important to add that animals develop distinctive colouration or other strange features for reasons other than the purpose

* In a just world, Zahavi's name would be much better known.

of sexual advertising to prospective mates. There is, for instance, 'aposematic colouration', which acts as a warning to predators not to eat or attack you. If you are a poisonous or foul-tasting beetle, for instance, it pays to look highly distinctive, so that birds will quickly learn to avoid eating you.† Lionfish (remember them?) deploy this tactic. Conversely, fruit (which is intended to be eaten) and flowers (which exist to attract the attention of insects) are highly distinctive, in order to encourage 'repeat visits'.

In a letter to Alfred Russel Wallace, Darwin wrote on 23 February 1867: 'On Monday evening I called on Bates and put a difficulty before him, which he could not answer, and as on some former similar occasion, his first suggestion was, "you had better ask Wallace." My difficulty is, why are caterpillars sometimes so beautifully and artistically coloured?' Darwin's theory of sexual selection, where distinctive colouration serves as a signal of sexual desirability, could not apply to caterpillars, since they are not sexually active until they metamorphose into butterflies or moths. Wallace replied the next day, suggesting that since some caterpillars 'are protected by a disagreeable taste or odour, it would be a positive advantage to them never to be mistaken for any of the palatable catterpillars [sic], because a slight wound such as would be caused by a peck of a bird's bill almost always I believe kills a growing catterpillar. Any gaudy and conspicuous colour therefore, that would plainly distinguish them from the brown and green eatable caterpillars, would enable birds to recognise them easily as at a kind not fit for food, and thus they would escape seizure, which is as bad as being eaten.'‡

† Ladybirds, for instance, secrete a foul-tasting chemical when eaten, and the brightly coloured dots on their backs advertise their inedibility.

‡ Before you ever dismiss someone who uses poor spelling and grammar, remember this paragraph. Wallace, one of the greatest minds in biology, left school at 14. In his famous 1858 paper to

Since Darwin was enthusiastic about the idea, Wallace asked the Entomological Society of London to test his hypothesis. The entomologist John Jenner Weir conducted experiments with caterpillars and birds in his aviary, and in 1869 provided the first experimental evidence for warning coloration in animals. The evolution of aposematism, literally a 'stay away sign' or 'warning off', surprised nineteenth-century naturalists because the conspicuous signal suggested a higher chance of predation. However, you might also argue that aposematic colouration might be explained as a form of costly signalling: 'I'm not trying to hide, therefore there might be a good reason not to eat me.'§

It might be a good rule of thumb for animals to avoid eating brightly coloured animals, since something that doesn't need to adopt camouflage has clearly survived through some strategy other than concealment, and hence it might be best avoided. Here again, we have a case where doing something ostensibly irrational conveys

---

the Linnean Society, he said of evolution that 'The action of this principle is exactly like that of the centrifugal governor of the steam engine, which checks and corrects any irregularities almost before they become evident; and in like manner no unbalanced deficiency in the animal kingdom can ever reach any conspicuous magnitude, because it would make itself felt at the very first step, by rendering existence difficult and extinction almost sure soon to follow.' The cybernetician Gregory Bateson observed in the 1970s that, though he saw it as an analogy, Wallace had 'probably said the most powerful thing that had been said in the nineteenth century'. In complex systems thinking, he understood the principle of self-regulating systems and feedback.

§ We might even apply the same reasoning to eighteenth-century British redcoats' red coats: 'I'm such a badass, I don't need to hide in the bushes like a Yankee.'

more meaning than something that makes sense. It has meaning precisely because it is difficult to do. It is not impossible to fake, but it is risky to do so – being highly visible but not poisonous is a mimicry strategy adopted by certain non-venomous snakes, for instance. What makes it risky is that, if any predator learns to tell you apart from the dangerous species you are imitating, he stands to make a killing – at your expense.

Wearing gold jewellery in South Central LA as a man is a doubly costly signal: it requires that you have the money to acquire the jewellery, but also conveys that you are hard enough to display it in public without fear of theft. I could afford to buy some fairly serious bling, but even on the sedate streets of London or Sevenoaks, I do not think, as a portly and out-of-shape middle-aged man, that I would have the necessary confidence to wear it.

3.10 NECESSARY WASTE

It was to explain his theory of sexual selection, and to defend his conception of the origin of species through natural processes rather than intelligent design, that Darwin wrote his second major book, *The Descent of Man*.* It broached a theory of sexual selection to explain, among other anomalies, how selection for fitness could produce such apparently fitness-reducing features as elaborate plumage.

The idea is simple, but not obvious. For a gene to persist, the body that carries it needs not only to survive but to reproduce – otherwise the gene will die out. Just as certain features such as acute vision or hearing and the capacity for swift movement will confer an advantage in survival, certain other features may confer an advantage in reproductive success – these are the attributes that allow you to mingle your genes either with larger numbers of mates or with mates that have a higher level of genetic quality. In humans and many other species, the emphasis placed on quality versus quantity may vary between the two sexes. In humans, females are naturally constrained in the quantity of offspring to which they can give birth, and so cannot obtain much advantage through mating indiscriminately; they need to consider other factors such as genetic

* The full title was *The Descent of Man, and Selection in Relation to Sex* (1871).

quality and the resources a man might be able to provide for his offspring.

But how should a female choose? Not being equipped with a gene sequencer, she relies on a mixture of sensory clues to spot the mating partners most likely to produce viable and successful offspring; age, size and resistance to parasites and illness may all be useful indicators. A creature that survives long enough to reach a great size or age clearly has what it takes to survive. Bullfrogs advertise their size and health by croaking, the deepness of the croak indicating size and its duration indicating fitness. Females which randomly developed a preference for deeper-throated, more persistently croaking frogs would on balance produce better-adapted offspring, since the trait was reliably correlated with quality. The two traits, deep-throatedness in males and a preference for it in females, will then grow in lockstep, since the genes for both will be increasingly found together.

There is a problem, however: what starts off as a reliable indicator of fitness can turn into an arms race. If you are a fit bullfrog, how long should you keep up your mating call? The only safe answer to this question is 'for a bit longer than any other bullfrog nearby'. As a result, a quality that starts off being prized as a useful proxy for fitness becomes exaggerated to an absurd degree, a process sometimes known as Fisherian runaway selection. In animals this can be extraordinarily wasteful. It seems that competition over antler size – which led to them ultimately growing to insane proportions – may have led to the extinction of the Irish elk.

The same competition can be just as damaging to humans when it manifests itself in unrestrained competition through extreme behaviour. Some academics have suggested that the Easter Island human civilisation might have been destroyed by competition between tribes over who could construct the largest and most numerous stone heads. There is no competition to build giant stone heads among modern humans,[†] but are car showrooms, DIY

† At least in my local area.

centres, women's clothes shops and shopping malls, or the preference for taking more expensive holidays, simply consumerist manifestations of the same uncontrolled competitive drive?

Of course, this competitive consumerism is not new. In 1759 Adam Smith made the following observation, in *The Theory of Moral Sentiments*:

'A watch, in the same manner, that falls behind above two minutes in a day, is despised by one curious in watches. He sells it perhaps for a couple of guineas, and purchases another at 50, which will not lose above a minute in a fortnight. The sole use of watches, however, is to tell us what o'clock it is, and to hinder us from breaking any engagement, or suffering any other inconveniency by our ignorance in that particular point. But the person so nice with regard to this machine, will not always be found either more scrupulously punctual than other men, or more anxiously concerned ... to know precisely what time of day it is. What interests him is not so much the attainment of this piece of knowledge, as the perfection of the machine which serves to attain it.'

Modern environmentalists also suggest that status-signalling competition between humans is destroying the planet. They propose that the Earth has enough resources to comfortably support the present population if we are all prepared to live modestly, but that natural rivalry can lead to ever-rising expectations – and with it increasing consumption. In many ways, this competition is not healthy – and nor does it necessarily contribute much to human happiness. In some ways it places people under an obligation to spend more money than they would otherwise choose to, just to maintain their status relative to other people.

There is an interesting debate to be had here between business and environmentalists. My contention is that, once you understand unconscious motivation, the widespread conviction that humans could be content to live without competing for status in an egalitarian state is nice in theory, but psychologically implausible.

Yet the status markers for which we compete don't have to be environmentally damaging; people can derive status from philanthropy as well as through selfish consumption. For instance, as Geoffrey Miller notes, a tribe where males advertised their hunting prowess by conspicuously sharing meat from their kills would prosper, as a result of economically irrational behaviour. On the other hand, an otherwise identical tribe whose males signalled their strength by violently fighting each other would suffer as a consequence: even the eventual winners of these contests might end up badly wounded and with a lower life expectancy. The first one is a positive sum game, while the other is anything but.‡ An extreme pessimist might suggest that, although competition for wealth markers is wasteful and harmful to the planet, it is a lot less harmful than many other forms of intergroup or interpersonal competition.§

Different forms of status seeking have effects on the wider populations that range from the highly beneficial to the downright disastrous. It has always struck me as odd that governments do not generally tax different forms of consumption at widely differing rates depending on their positive or negative externalities (as they do with tobacco, alcohol and petrol). I am, as you might expect from my job, fairly forgiving of most forms of consumerism, but there are some activities, such as the mining of diamonds for jewellery, which seem entirely without merit. I may be alone in saying this, but I don't think evolution by natural selection was Darwin's most interesting idea. Earlier thinkers, from Lucretius to Patrick Matthew, had also recognised the basics of natural selection, and many practical people, whether pigeon fanciers or dog breeders, had also grasped the essential principles. Had neither Darwin or Wallace existed, it seems inevitable that someone else would have come up with a similar theory.

However, the theory of sexual selection was a truly extraordinary, outside-the-box idea, and it still is; once you understand it, a whole

‡ The Russian oligarchy seems to manifest certain features of this rivalrous second kind of tribe.

§ For instance, it is arguably better for mildly sociopathic males to aspire to own a large yacht than to aspire to run the secret police.

host of behaviours that were previously baffling or seemingly irrational suddenly make perfect sense. The ideas that emerge from sexual selection theory explain not only natural anomalies such as the peacock's tail, but also the popularity of many seemingly insane human behaviours and tastes, from the existence of Veblen goods¶ such as caviar, to more mundane absurdities such as the typewriter.

For almost a century in which few men knew how to type, the typewriter must surely have damaged business productivity to an astounding degree, because it meant that every single communication in business or government had to be written twice: once in longhand by the originator and then again by the typist or typing pool. A series of simple amendments could delay a letter or memo by a week, but the ownership and use of a typewriter was a signal that you were a serious business – any provincial solicitor who persisted in writing letters by hand became a tailless peacock.

Take note that I have committed the same offence that everyone else does when writing about sexual selection: I have confined my examples to those occasions where it runs out of control and leads to costly inefficiencies, such as typewriters, Ferraris and peacocks tails.** This is unfair.

In the early stages of any significant innovation, there may be an awkward stage where the new product is no better than what it is seeking to replace. For instance, early cars were in most respect worse than horses. Early aircraft were insanely dangerous. Early washing machines were unreliable. The appeal of these products was based on their status as much as their utility.

¶ Goods that increase in demand as their price goes up.

** You may have noticed that there are very few famous Belgians – this is because when you are a famous Belgian (like Magritte, Simenon or Brel) everyone assumes you are French. In the same way, there are few commonly cited examples of successful sexual selection: when sexual selection succeeds, people casually attribute the success to natural selection.

The tension between sexual and natural selection – and the interplay between them – may be the really big story here. Many innovations would not have got off the ground without the human instinct for status-signalling,†† so might it be the same in nature? In other words, as Geoffrey Miller says, might sexual selection provide the 'early stage funding' for nature's best experiments? For example, might the sexual signalling advantages of displaying an increasing amount of plumage on a bird's sides‡‡ have made it possible for them to fly? The human brain's capacity to handle a vast vocabulary may have arisen more for the purposes of seduction than anything else – but it also made it possible for you to read this sentence. Most people will avoid giving credit to sexual selection where they possibly can because, when it works, sexual selection is called natural selection.

Why is there a reluctance to accept that life is not just a narrow pursuit of greater efficiency and that there is room for opulence and display as well? Yes, costly signalling can lead to economic inefficiency, but at the same time this inefficiency establishes valuable social qualities such as trustworthiness and commitment – politeness and good manners are costly signalling in a face-to-face form. Why are people happy with the idea that nature has an accounting function, but much less comfortable with the idea that it also has a marketing function? Should we despise flowers because they are less efficient than grasses? Even Darwin's great contemporary and collaborator Wallace hated the idea of sexual selection; for some reason, it sits in the category of ideas that most people – and especially intellectuals – simply do not want to believe.

†† For a good decade or so, cars were inferior to horses as a mode of transport – it was human neophilia and status-seeking, rather than the pursuit of 'utility', which gave birth to the Ford Motor Company – Henry was a bit of 'boy racer' in his youth.

‡‡ Rather than, like the peacock, senselessly overinvesting in the rear spoiler.

3.11 ON THE IMPORTANCE OF IDENTITY

Remember that without distinctiveness, mutualism of the kind found in bees and flowers cannot work, because an improvement in a flower's product quality would not result in a corresponding increase in the bees' loyalty. Without identity and the resulting differentiation, a breed of flower would give away extra nectar for no gain, as the next time, the bees would simply visit the less-generous-but-identical-looking flower next to it. Over time, flowers would end up in a 'race to the bottom', producing as little costly nectar as possible and relying on their similar appearance to other, more generous flowers to preserve the bees' supply of nectar and to maintain the incentive for them to continue travelling from plant to plant.

We need to consider whether the same process occurs in business, as well as in nature. Are brands essential to making capitalism work?

3.12 HOVERBOARDS AND CHOCOLATE: WHY DISTINCTIVENESS MATTERS

Many of you reading this may be too young to remember fighting over Cabbage Patch dolls or Buzz Lightyear figures, but let us pause to recall what wasn't *quite* the hot craze of Christmas 2015, because it provides a valuable lesson in the wider economic importance of brands.

By this, I mean the Hover Board. Or the Hoverboard, the Swagway, the Soarboard, the PhunkeeDuck or the Airboard. There was never was an agreed name for these things, because they were sourced from a range of manufacturers in Shenzhen and named by the local distributors rather than the makers. The idea had not been commissioned by one large company but rather seemed to arise from experimentation: these unusual origins provide us with a rare and unusual test case of what happens to innovation in the absence of brands.

The board is an interesting product, and I'm sure many of you may instinctively have wanted to try or buy one, but you didn't, did you? First of all, you didn't know which one to buy – some had lights or Bluetooth speakers,* some had larger wheels and some had a higher or lower price. And in the absence of recognisable brands, it was impossible to make sense of the category – as neuroscientists

* Why?

What this product needs is a brand. Without a distinctive brand identity, there is no incentive to improve your product – and no way for customers to choose well, or to reward the best manufacturer.

have observed, we don't so much choose brands as use them to aid choice. And when a choice baffles us, we take the safe default option – which is to do nothing at all.

Secondly, we felt uneasy buying something that cost a few hundred pounds without the reassurance of a recognisable name. British ad-man Robin Wight calls this instinct 'the Reputation Reflex' – although instinctive and largely unconscious, it is perfectly rational, because we intuitively understand that someone with a reputable brand identity has more to lose from selling a bad product than someone with no reputation at risk.† Finally, while we were still wondering whether to take the plunge, news spread that several boards had burst into flames while charging, in one case setting fire to a house. The problem was confined to a few makes, but without knowing which specific brands to avoid, it sullied the whole category.

† Had there been a Samsung, LG or Dyson hoverboard on offer, you very well might have bought one.

Without the brand feedback mechanism, there was no incentive for any one manufacturer to make a safer, better version of the board, since they were not positioned to reap the gains. As a result, the market became a commoditised race to the bottom, in which both innovation and quality control fail. Why make a better product if no one knows it was you who made it? So no one did make a better board, and the whole category more or less died as a result. It may correct itself if better boards arise, or if a shrewd company such as Samsung cannily attaches its name to the best. Noticeably, brands such as Juul and Vype are starting to emerge in the similarly haphazard vaping market.

In many ways, expensive advertising and brands arise as a solution to a problem identified by George Akerlof in his 1970 paper 'The Market for Lemons' in the *Quarterly Journal of Economics*. The problem is known as 'information asymmetry', whereby the seller knows more about what he is selling than the buyer knows about what he is buying. This lesson was learned the hard way in Eastern Bloc countries under communism; brands were considered un-Marxist, so bread was simply labelled 'bread'. Customers had no idea who had made it or whom to blame if it arrived full of maggots, and couldn't avoid that make in future if it did, because all bread packaging looked the same. Unhappy customers had no threat of sanction; happy customers had no prospect of rewarding producers through repeat custom. And so the bread was rubbish.

The production of rivets under communism demonstrated a similar pattern. Typically a factory was given a monthly quota that it was required to manufacture – the unbranded rivets were then sent to a central rivets depository, where they intermingled with all the other factories' rivets. From there, all the rivets, whose provenance was by now completely indistinguishable, would be transported to wherever they were needed. The Soviets soon found that, without a maker's name attached to a product, no one had any incentive to make a quality product, which pushed quantity upwards and quality downwards. The easiest way to produce a million rivets every month was to produce a million bad rivets, which soon led to ships falling apart. Furthermore, you did not know which factory to blame, because the rivets had become commoditised, which is to say anonymised. Eventually the regime swallowed their ideological pride and made factories stamp their names on their rivets – the feedback mechanism was restored and quality returned to acceptable levels.

I recently met a woman who lived under communism in Romania. A popular chocolate bar in that country at the time was manufactured in three different factories, but the quality of production between them differed so widely that they effectively produced three entirely different qualities of chocolate under the same brand name. By folding back part of the wrapper, it was possible to see an alphanumeric code that denoted, presumably for reasons of safety, which of the three factories had produced that particular bar. My friend, a young girl at the time, was under strict instructions from her mother only to buy this chocolate bar if the letter 'B' was displayed under the fold in the wrapper; if either of the other two letters were displayed, she wasn't to buy it at all.

Without the feedback loop made possible by distinctive and distinguishable petals or brands, nothing can improve. The loop exists because insects or people learn to differentiate between the more and less rewarding plants or brands, and then direct their behaviour accordingly. Without this mechanism there is no incentive to improve your product, because the benefits will accrue to everybody equally; in addition, there is an ever-present incentive to let product quality slip, because you will reap the immediate gains, while the reputational consequences will hurt everyone else equally. This explains why it is necessary for markets to endure the apparent inefficiency of supporting different and competing products with expensively differentiated identities, in order to reward quality control and innovation.

Several years ago there was a national crisis in Britain regarding confidence in the meat supply, after horsemeat was found to have been secretly mixed into supplies of certified beef.‡ While nobody

‡ In truth, had we been French, this wouldn't have been much of a crisis, but to the British, the idea of eating horse is anathema.

died – in fact, nobody even got ill – it nevertheless significantly eroded the public's trust in the food industry, and rightly so. It wasn't branded beef that was affected – McDonald's emerged from the scandal completely unscathed – the beef that was contaminated was typically labelled 'certified beef from a variety of sources'. No one supplying beef that they knew would be mingled with other people's beef had any fear of reputational shame, and there was consequently nothing to discourage any of the suppliers of this commoditised beef from adding a bit of horsemeat into the mix.

This matters, because conversations about the marketing of brands tend to focus on hair-splitting distinctions between fairly good products. We often forget that, without this assurance of quality, there simply isn't enough trust for markets to function at all, which means that perfectly good ideas can fail.

Branding isn't just something to add to great products – it's essential to their existence.

Evolution solved the problem of asymmetric information and trust for flowers and bees back when our ancestors were still living in trees. Bees have been around for at least 20 million years, floral plants a good deal longer. My analogy between signalling in the biological world and advertising in the commercial world may explain something I have noticed for years: if you talk to economists, they tend to hate advertising and barely understand it at all, while if you talk to biologists they understand it perfectly. For decades, the most sympathetic ear I had at *The Economist* in London was not their marketing correspondent (who seemed to genuinely hate marketing) but their science correspondent, whose background was as an evolutionary biologist.

PART 4: SUBCONSCIOUS HACKING: SIGNALLING TO OURSELVES

4.1 THE PLACEBO EFFECT

I earlier described how it is often necessary to use oblique approaches to change the behaviour of others. Now I would like to suggest it can be equally necessary to use similar approaches to change your *own* behaviour.

We'll start by looking at the power of the placebo. My grandfather was a doctor from 1922 to the mid-1950s. He claimed that it was only after the advent of penicillin that he became a true medic: before antibiotics, he was partly a glorified witch doctor – the support he offered his patients through the psychological value of a doctor's visit being as important as the pharmaceutical value of anything he prescribed.

Are placebos, or placebo treatments such as homeopathy, scientific? Well, yes and no. And do they help? Well, sometimes. Placebos have no direct medical efficacy, but their effect on our psychology may be just as significant as a medical effect in some cases, especially if the condition – chronic pain, say, or depression – is more psychological than physiological.*

* In the same way, the psychological solutions I propose in this book concern those social or commercial issues which are more psychological not physical. A famine will not be cured by psychological intervention – but over-eating might be.

I make a very simple point here: the fact that something does not work through a known and logical mechanism should not make us unwilling to adopt it. We used aspirin to reduce pain for a century without having the faintest idea of why it worked. Had we believed it was made from the tears of unicorns, it would have been silly, but it wouldn't have made the product any less effective.

4.2 WHY ASPIRIN SHOULD BE REASSURINGLY EXPENSIVE

A few years ago, the rationalist killjoys at the Australian Competition and Consumer Commission prosecuted the global consumer goods manufacturer Reckitt Benckiser over four products: Nurofen Migraine Pain, Nurofen Tension Headache, Nurofen Period Pain and Nurofen Back Pain. Their complaint was that 'each product claimed to target a specific pain, when in fact it was found that they all contained the same amount of the same active ingredient, ibuprofen lysine' – the problem was that these variants were often sold at a higher price than the basic brand, despite being pharmacologically identical.

While I am sure the ACCC's chemical facts were accurate, their psychology seems to have been wrong because, for me, Nurofen hadn't gone far enough. I want to see even more specific variants of pain relief: 'I Can't Find My Car Keys Nurofen' or 'Nurofen for People Whose Neighbours Like Reggae'. Again, these need contain no additional ingredients: the only distinguishing feature would be the packaging and the promise. I'm not being entirely frivolous: research into the placebo effect shows that branded analgesics are more effective. Furthermore, promoting a drug as a cure for a narrowly defined condition, as Nurofen did, also increases placebo power, as does raising its price or changing the colour: everything the company was doing added to the efficacy of the product.

It is impossible to buy expensive aspirin in the UK, yet it is a waste of this wonder drug to sell it for 79p in drab packaging, when you could make it much better by packaging it lavishly, colouring the pills red* and charging more. Sometimes I have a £3.29 headache rather than a 79p one. I try to stockpile the pricier brands I buy in the US, because I find they work better.

Yes, I know it's bullshit but, as we've already seen, placebos work even if you tell people they are placebos. Or, to put it another way, a dock leaf might soothe the pain of a nettle sting on even Richard Dawkins's leg, regardless of any scientific evidence he had of their uselessness.

The psychologist Nicholas Humphrey argues that placebos work by prompting the body to invest more resources in its recovery.† He believes that evolution has calibrated our immune system to suit a harsher environment than the current one, so we need to convince our unconscious that the conditions for recovery are especially favourable in order for our immune system to work at full tilt. The assistance of doctors (whether witch or NHS), exotic potions (whether homeopathic or antibiotic) or the caring presence of relatives and friends can all create this illusion, yet policymakers hate the idea of any solution that involves such unconscious processes – too little is spent on researching the placebo effect in proportion to its importance.‡

Understanding the placebo effect is a useful way to begin to understand other forms of unconscious influence; it explains why we often behave in apparently irrational ways in order to influence unconscious processes – both our own and those of others. Additionally, our reluctance to exploit the placebo effect may offer some clues about our wider reluctance to adopt psychological solutions to problems, particularly when they are slightly counterintuitive or not conventionally logical. Let me explain.

---

* Painkillers are more effective when the pills are red.

† His writings on the subject include 'The Placebo Effect' in R. L. Gregory (ed.), *Oxford Companion to the Mind* (2004).

‡ If you suggested the NHS invested in more elaborate drugs packaging, they'd have palpitations.

The placebo effect, like many other forms of alchemy, is an attempt to influence the mind or body's automatic processes. Our unconscious, specifically our 'adaptive unconscious' as psychologist Timothy Wilson calls it in *Strangers to Ourselves* (2002), does not notice or process information in the same way we do consciously, and does not speak the same language that our consciousness does, but it holds the reins when it comes to much of our behaviour. This means that we often cannot alter subconscious processes through a direct logical act of will – we instead have to tinker with those things we *can* control to influence those things we *can't* or manipulate our environment to create conditions conducive to an emotional state which we cannot will into being.

Think about it. There are some phrases that just wouldn't appear in the English language:[§]

1. 'I chose not to be angry.'
2. 'He plans to fall in love at 4.30pm tomorrow.'
3. 'She decided that she was no longer to feel uneasy in his presence.'
4. 'From that moment on, she determined no longer to be afraid of heights.'[¶]
5. 'He decided to like spiders and snakes.'

§ Or any other language, come to that.

¶ I have a friend who is not only terrified of heights – she is also terrified of tomatoes. I myself, like the late Steve Jobs, suffer from koumpounophobia, the fear of buttons. Mine is a mild case: I am now, as an adult, happy to wear buttons when they are firmly attached to clothing, but find them disquieting when loose. Steve's phobia was more severe – he would never wear anything with visible buttons. Some theorise that this influenced his design philosophy, driving his refusal to produce a phone until you could produce one without a button-containing keyboard.

Things like this are not under our direct control, but are rather the product of instinctive and automatic emotions. There is a good evolutionary reason why we are imbued with these strong, involuntary feelings: feelings can be inherited, whereas reasons have to be taught, which means that evolution can select for emotions much more reliably than for reasons. To ensure your survival, it is much more reliable for evolution to give you an instinctive fear of snakes at birth than relying on each generation to teach its offspring to avoid them. Things like this aren't in our software – they are in our hardware.

In the same way, we all accept the fact that there are large areas of bodily function that we cannot control directly: I cannot make my pupils contract or dilate at will, and nor can I increase or slow my heart rate by telling my heart to speed up or slow down; that's to say nothing about other bodily functions, such as digestion, sexual arousal, the secretions of the pancreas, the actions of the endocrine system or the workings of the immune system. For perfectly sensible evolutionary reasons, the regulation of these functions does not impinge on consciousness.** You might like to think of these processes as operating on an 'auto' setting, similar to that which exists on a modern camera – the useful function that means you don't have to spend time fiddling around with the aperture, focus and shutter-speed settings every time you want to take a half-decent photograph.

** It would be a bit odd if they did: 'Just hold on a second, darling, will you? I'm just elevating my testosterone settings and turning my tumescence levels up to eight.'

4.3 HOW WE CAN 'HACK' WHAT WE CAN'T CONTROL

As with an automatic camera, so with your body's autonomic systems – you can't directly control either of them, but you can 'hack' them obliquely, by deliberately contriving the conditions that will generate the automatic response you want. To continue the photography analogy, imagine that you have a fully automatic camera and wish to deliberately overexpose a photograph. There isn't a dial that enables you to prolong the shutter speed or enlarge the aperture, but you can achieve the same effect by pointing at something dark, triggering the auto-exposure mechanism and then panning back to the better-lit subject of your photograph.

I have always been – unfashionably for a European – a devotee of automatic transmission in cars* and, as anyone who has driven the same model of automatic car for any length of time knows, you soon learn how to induce or prevent a gear change using the accelerator pedal alone. You do this by becoming increasingly attuned

* Purist, petrolhead (especially German) friends have always ridiculed me for this. 'Ja, but you don't have ze same sense of kontrol,' they say. This is now rubbish – but, in defence of my Teutonic friends, European automatics were often quite bad 30 years ago.

to the behaviour of your automatic gearbox, unconsciously developing the skill of encouraging it to do what you want. On nearing the brow of a hill, for instance, you might instinctively take your foot off the accelerator to prevent the gearbox from changing down unnecessarily for the remaining short climb. Manual car devotees are blind to this skill, because it is something you only learn to do through repeatedly driving the same automatic car. The truth is that you *can* control the gearbox of an automatic car, but you just have to do it obliquely. The same applies to human free will: we can control our actions and emotions to some extent, but we cannot do so directly, so we have to learn to do it indirectly – by foot rather than by hand.

This indirect process of influence applies to all complex systems, of which the automatic gearbox and human psychology are merely two examples.† The problems we face arise because policy problems are given to the intellectual equivalent of manual car drivers, who believe that the only acceptable way to change gear is to use a gearstick, rather than indirectly with an accelerator pedal. But the trick is to accept that driving an automatic is much more creative than driving a manual: with a manual you merely tell the gearbox what to do, but when you are driving an automatic, you have to use seduction.‡

Imagine that you wish to dilate your pupils, increase your heart rate, decrease your heart rate or boost your immune system. Again, you cannot do this by a direct act of will, but you *can* use conscious mechanisms to produce unconscious effects. For instance, you can contract your pupils by staring at a light bulb or dilate them by walking into a darkened room.§ You can increase your heart rate through jogging or decrease it through the practice of yoga or

---

† The steam regulator (remember Alfred Russel Wallace, from Chapter 3.9) is, of course, another.

‡ British readers, I'm aware that I'm at risk of sounding a bit 'Swiss Tony' here.

§ Or by looking at pornography – apparently.

meditation. And, yes, if Nicholas Humphrey is right, you might be able to boost your immune system in the same way – you simply have to create the conditions that lead your immune system to believe that the present is a particularly advantageous time to invest its resources in healing wounds or combating infection. The actions required to create such conditions may involve a certain degree of what appears to be bullshit – but it is only bullshit when you don't know what its reason is.

It is this oblique hacking of unconscious emotional and physiological mechanisms that often causes suspicion of the placebo effect, and of related forms of alchemy. Essentially we like to imagine we have more free will than we really do, which means we favour direct interventions that preserve our inner delusion of personal autonomy, over oblique interventions that seem less logical.

4.4 'THE CONSCIOUS MIND THINKS IT'S THE OVAL OFFICE, WHEN IN REALITY IT'S THE PRESS OFFICE'

Our conscious mind tries hard to preserve the illusion that it deliberately chose every action you have ever taken; in reality, in many of these decisions it was a bystander at best, and much of the time it did not even notice the decision being made. Despite this, it will still construct a story in which it was the decisive actor. For instance, 'I saw the bus coming and jumped back on the kerb,' while in fact, you may well have started jumping before you were even consciously aware of the bus.

In the words of Jonathan Haidt,* 'The conscious mind thinks it's the Oval Office, when in reality it's the press office.' By this he means that we believe we are issuing executive orders, while most of the time we are actually engaged in hastily constructing plausible post-rationalisations to explain decisions taken somewhere else, for reasons we do not understand. But the fact that we can deploy reason to explain our actions post-hoc does not mean that it was reason that decided on that action in the first place, or indeed that the use of reason can help obtain it.

Imagine an alien species with the power to fall asleep at will – they would regard human bedtime behaviour as essentially ridiculous. 'Rather than just going to sleep, they go through a

* In *The Righteous Mind* (2012).

strange religious ritual,' an alien anthropologist would remark. 'They turn off lights, reduce all noise to a minimum and then remove the seven decorative cushions which for no apparent reason are placed at the head of the bed.† Then they lie in silence and darkness, in the hope that sleep descends upon them. And rather than simply waking up when they wish, they program a strange machine which sounds a bell at an appointed time, to nudge them back into consciousness. This seems ridiculous.' Similarly, imagine an alien species that could decide how happy it wanted to be. They would regard the entire human entertainment industry as a spectacular economic waste.

We don't pretend that we can sleep at will or control our levels of contentment, but much of the time we pretend that conscious human agency is the only force that drives our behaviour, and therefore disparage other less obvious behaviours that we have adopted to hack our unconscious processes as if they were irrational, wasteful or absurd. This leads to the same frustration felt by a manual car driver in control of an automatic car for the first time. Someone who has not mastered the technique of oblique influence can only envisage direct intervention to achieve the desired effect, as follows:

1. If you want to change gear, you must move the gearstick.
2. If you want people to work harder, you must pay them more.
3. If you want people to give up smoking, you must tell them it kills them.
4. If you want people to take out pensions, you must give them a tax incentive.
5. If you want people to pay more for your product, you must make it objectively better.

† I'm with the aliens here. Does anyone else have to put up with this? It drives me nuts.

6. If you want to improve a train journey, you must make the trains faster.
7. If you want to improve your well-being, you must consume more resources.
8. If you want people to get better, you must give them an active drug.

I owe my explanation of the placebo theory to its author, Nicholas Humphrey. To me, his theory seems to be among the most significant theories in the field of psychology. Indeed, given its potential value to human health, it is inexplicable to me that more use has not been made of it, or at the very least that it has not been more widely investigated. It could potentially change the whole practice of medicine, but I suspect that the reason people seem less than eager to pursue the implications of Humphrey's ideas is because it contains a whiff of alchemy.

An article in *New Scientist* in 2012 examining the nature of the placebo effect described new evidence from a model that offered a possible evolutionary explanation. It suggested that the immune system has 'an on-off switch controlled by the mind', an idea first proposed by psychologist Nicholas Humphrey a decade or so earlier.

Pete Trimmer, a biologist at the University of Bristol, observed that the ability of Siberian hamsters to fight infections varied according to the lighting above their cages – longer hours of light (mimicking summer days) triggered a stronger immune response. Trimmer's explanation was that the immune system is costly to run, and so as long as an infection is not lethal, it will wait for a signal that fighting it will not endanger the animal in other ways. It seems that the Siberian hamster subconsciously fights infection more energetically in summer because that is when food supplies are sufficiently plentiful to sustain an immune response. Trimmer's model demonstrated that in challenging environments, animals fared better by weathering infections and conserving resources.

Humphrey argues that people subconsciously respond to a sham treatment because it assures us that it will weaken the infection without overburdening the body's resources. In populations where food is plentiful we can, in theory, mount a full immune response at any time, but Humphrey believes that the subconscious switch has not yet adapted to this – thus it takes a placebo to convince the mind that it is the right time for an immune response.

4.5 HOW PLACEBOS HELP US RECALIBRATE FOR MORE BENIGN CONDITIONS

It's interesting that Humphrey suggests that our body's immune system is calibrated to suit a much tougher environment than the one in which we find ourselves.* My parents' generation lived through the food shortages during the Second World War and the long period of rationing which followed. My aunt, late in life, could still not bring herself to throw food away uneaten, even when the contents of her fridge had decayed to the point of becoming a biohazard – her attitude to waste had been calibrated during a time of great scarcity.

In the same way, the human immune system has over time been calibrated to promote survival in conditions far harsher than those of today. Previously, you could not risk squandering resources too hastily when there was an ever-present risk that you might starve to death, freeze to death or be immobilised by the body's immune response.† To recalibrate our immune response to levels

* A similar explanation is sometimes used to explain human obesity – for most of human evolution, a reliable instinct was 'If you find anything tasty, eat a lot of it.'

† Many of the unpleasant symptoms of an illness – a high temperature, say – arise not from the illness itself, but from the body's attempts to fight it.

appropriate to the more benign conditions we experience in everyday modern life, it may be necessary to deploy some benign bullshit.‡ This, I suppose, was what my grandfather was doing in the days before antibiotics, when he cheered up his patients with banter and encouraged them to wrap up warmly, stay in bed, feed themselves well and drink medicinal whiskies – perhaps prescribing for good measure some ineffectual pills that nonetheless created enough of an illusion of optimism for the patient's body to enter 'healing mode'.

When I met Nicholas Humphrey at an Indian restaurant in London,§ he had by this point expanded his theory beyond health and the immune system: as well as hacking the immune system, he believes that humans regularly deployed oblique methods to generate bodily states and emotions which, like our immune response, we cannot consciously will into action – but which we can coax into existence. In particular, he mentioned bravery placebos, devices designed to achieve higher levels of bravery than could be obtained through conscious will alone.

Think about this for a moment. Bravery is, for most people, not a consciously determined state – it's automatic, not manual.¶ Although your mother might have exhorted you when going to primary school to 'Be brave', there is honestly little we can do to effect this condition

‡ If that means homeopathy, so be it.

§ For those interested in learning more about this theory, there is a very rewarding YouTube video in the 'enemies of reason' series in which Nicholas Humphrey argues his case against Richard Dawkins, the high priest of reductionist rationality.

¶ Special forces personnel may be exceptions here – they may be selected for their ability to switch off fear, and some show strong signs of psychopathy. Nevertheless, for the rest of us grunts, whether to be frightened or not isn't a matter in which we have much choice.

in ourselves, any more than we can 'decide to go to sleep'. So, as Humphrey explained, much of the paraphernalia and practice of the military – flags, drums, uniforms, square-bashing, regalia, mascots and so forth – might be effectively bravery placebos, environmental cues designed to foster bravery and solidarity.

As with getting to sleep, the trick in generating bravery lies in consciously creating the conditions conducive to the emotional state. With sleep this might mean fluffy pillows, darkness and silence;** in the case of bravery it might involve trumpets, drums, banners, uniforms, camaraderie£ and so on. Soldiers live together, call themselves 'brothers in arms', march in lockstep, wear the same clothes and are allocated to 'fictive-kin' groups such as platoons, regiments and divisions – much of this fosters the illusion that you might be willing to make the ultimate sacrifice for the people within your group.

This theory makes sense of much behaviour that initially appears absurd – it is an idea which, once entertained, stays in your mind for years and helps you to view people's actions in an entirely new light. The strangest aspect of it is that we all spend a considerable amount of time and money essentially signalling to *ourselves*: many of the things we do are not be intended to advertise anything about ourselves to others – we are, in effect, advertising to ourselves.†† The evolutionary psychologist Jonathan Haidt refers to such activities as 'self-placebbing.' Once we understand the concept, a great deal of bizarre consumerism will make more sense.

** And/or a quarter of bottle of whisky.

†† When we buy L'Oréal products, perhaps we are advertising to ourselves 'that I'm worth it'.

4.6 THE HIDDEN PURPOSES BEHIND OUR BEHAVIOUR: WHY WE BUY CLOTHES, FLOWERS OR YACHTS

There is an important lesson in evaluating human behaviour: never denigrate a behaviour as irrational until you have considered what purpose it *really* serves. It is clearly irrational to buy a £250,000 Ferrari as a form of everyday transportation, when a perfectly serviceable car can be bought for a fraction of the price. On the other hand, as an aphrodisiac or as a means of humiliating a business rival, it does rather outclass a Honda Civic. I'm not into Ferraris,* but I can understand that they probably confer a kind of confidence on the driver.

As an interesting thought experiment, I often construct fake advertising slogans for various products – in particular the slogans that they would adopt if they could afford to be completely honest about why people bought them. 'Flowers – the inexpensive alternative to prostitution' – that kind of thing. They are rather like the slogans that appear throughout the film *The Invention of Lying* (2009),† which is set in a world in which everyone initially tells the truth about everything. 'Pepsi – for when they don't have Coke'.‡

---

* Though I might buy one if I got divorced!

† A film where, to be completely honest, the premise is more interesting than the execution.

‡ Occasionally, ads playing to this idea have run. 'Small penis? Have we got a car for you!' said one Canadian

I invent my brutally honest slogans to make the point that most products have both an ostensible, 'official' function and an ulterior function. The main value of a dishwasher, I would argue, is *not* that it washes dirty dishes, but that it provides you with an out-of-sight place to put them. The main value of having a swimming pool at home is not that you swim in it, but that it allows you to walk around your garden in a bathing costume without feeling like an idiot. A friend who had been invited to spend a week on a luxury yacht explained why they are so popular with megalomaniacs: 'You can invite your friends to join you on holiday, then spend the week treating them like you are Captain Bligh.' If you have the most magnificent villa in the world, there is still the risk that your friends and rivals might hire a car and wander off on their own: on a megayacht, however, they are your captives.[§]

One problem (of many) with Soviet-style command economies is that they can only work if people know what they want and need, and can define and express their wants adequately. But this is impossible, because not only do people not know what they want, they don't even know why they like the things they buy. The only way you can discover what people really want (their 'revealed preferences', in economic parlance) is through seeing what they actually pay for under a variety of different conditions, in a variety of contexts. This requires trial and error – which requires competitive markets and marketing.

---

advert for a Porsche dealership (before I imagine Porsche stripped it of its franchise).

§ If you are not a megalomaniac, do not buy a megayacht. A friend of mine organised the sale of them for many years: he said that the main lesson he learned about yachts is that, above a certain minimum, the pleasure they provide is in inverse proportion to their size. Also, a very large yacht can only be moored in a few harbours, meaning that you will often end up moored next to a yacht even larger than your own.

The intriguing thing about Uber as an innovation was that no one really asked for it before it existed.¶ Its success lay in a couple of astute psychological hacks: the fact that no money changes hands during a trip is one of the most powerful – it makes using it feel like a service rather than a transaction.**

Take the control panels in elevators. One of the buttons found on them, the 'door close' button, is quite interesting, because on many (and perhaps most) elevators, it is actually a placebo button – it is connected to nothing at all. It is there simply to make impatient people feel better by giving them something to do and the illusion of control. It is, in effect, a civilised alternative to a punchbag. I don't know if this is a bad thing – it's definitely a lie, but perhaps it's a white lie – one whose sole job is to make someone feel better. Since the only possible purpose of a 'door close' button is to make impatient people relax, perhaps it makes no difference whether it achieves this end through mental or mechanical means.††

The use of placebo buttons is more common than we realise. Many pedestrian crossings have buttons that also have no effect at all – the traffic lights are set to a timed sequence.‡‡ However, here

---

¶ Actually, I did, but that was because I had spent some years being obsessed with the effects of uncertainty on waiting. When I shared my suggestion with other people, they mostly shrugged.

** Additionally, we also experience credit card transactions as being over 15 per cent cheaper than equivalent cash transactions.

†† I am not proposing the use of placebo buttons in aircraft cockpits – though in fly-by-wire aircraft something similar exists, where the on-board computer interprets the pilot's intentions rather than obeying his instructions directly.

‡‡ Pizza delivery apps often give you step-by-step updates on the creation, baking, quality control, boxing and delivery of your pizza. How much of this is actually true real-time information, and how much is

the presence of the button is a rather more benign lie: how many fewer people would wait for the green man if there were no button to press? And how many more people would wait for the green man if there were a digital display of the seconds to wait before its appearance? In countries including Korea and China, accidents at intersections have been reduced by simply displaying the number of seconds remaining before the lights turn green.[§§]

This is because the mammalian brain has a deep-set preference for control and certainty. The single best investment ever made by the London Underground in terms of increasing passenger satisfaction was not to do with money spent on faster, more frequent trains – it was the addition of dot matrix displays on platforms to inform travellers of the time outstanding before the next train arrived.

Let's apply this insight to something more momentous. If we know that people hate uncertainty, and that men are disproportionately reluctant to undergo medical testing, how can we combine the two insights and come up with a solution? What if the reason men hate undergoing certain tests is that they are unconsciously averse to the uncertainty they experience while waiting for the results? They can't tell us this, because they don't know – remember the lenses in the broken binoculars? Logic won't tell us this, either, but we can test – by seeing what happens if we make a promise: 'If you have this test, we will text you the results within 24 hours.' To date, nobody has thought that kind of promise might

simply there to give the reassuring impression of progress? I am fairly sceptical, but I kind of like the illusion of progress, nonetheless. What these functions say is, 'Relax – we haven't forgotten about you.'

§§ In Korea they even tested the obverse idea – of a digital display on green traffic lights to show oncoming drivers the number of seconds left before the light turned red. This, as you will realise with just a moment's thought, was a very bad idea indeed.

be relevant: nobody considered that the uncertain delay between having the test and getting the result might influence the human propensity to undergo the test in the first place.

Credit card companies have discovered this already, with promises like 'Apply now and get approval within 12 hours' – they found, through testing, accident or experimentation, that this made a difference to people's keenness to respond. Whether you're carrying out market research or using neoclassical economic assumptions, you wouldn't realise that the amount of time spent in uncertainty might be an important factor.

A simple thought experiment might help here. If there were a medical device whereby you could press a button and get an immediate reading with an early warning of prostate cancer, I suspect most men would been keen on the idea. By contrast, our willingness to book an appointment, meet a phlebotomist and wait two weeks for a result is very low.

4.7 ON SELF-PLACEBBING

As we've seen, one way to understand military paraphernalia of uniforms, trumpets, drums and regalia is to consider its value as a 'bravery placebo'. But what are the other emotions that we might wish to hack through the use of a similar 'Humphrey placebo'? Two spring to mind immediately: the need to inspire confidence in ourselves, and the need to inspire trust in others.

I have twin 17-year-old daughters and I love them dearly, except when the time comes to leave the house. Their cosmetic regime is beyond ridiculous: before attending any party or function, it isn't unusual for them to spend an hour and a half getting ready, most of which is dedicated to face painting or eyebrow tweaking of one kind or another. I find it irritating enough to have to shave each morning; how they can put up with this rigmarole is completely beyond me. Evolutionary psychology might explain my daughters' infuriating behaviour in several ways: they might be trying to enhance their appearance and reproductive fitness by signalling to the opposite sex. They might be seeking to maintain or enhance their status among their own sex. Or they might be doing it to signal to themselves.

Whichever is right (and they are not mutually exclusive), my offspring are clearly are not alone in their behaviour. I once attended a presentation on the worldwide beauty industry, which includes clothing, perfume and cosmetics. I was briefly confused by a chart that showed billions of US dollars as its scale but which listed annual spending figures in thousands, before realising that the annual expenditure we were discussing ran into *trillions* of US dollars. It turns out that more is spent on female beauty than on education.*

Once you understand the placebo, I think you'll agree that a large part of the two trillion dollars spent on female self-beautification is not spent in order to appeal to the opposite sex; to put it bluntly, as a woman, it simply isn't that difficult to dress in a way that appeals to men – you just have to wear very little.† There are also some trends in female fashion, high-waisted trousers, for instance, which men find fairly repellent.‡ It seems likely that a significant part of what you're doing when you spend two hours on self-grooming is self-administering a confidence placebo to produce emotions that you can't generate through a conscious act of will.§

Men have equivalent placebo vices, of course: of these, a love of cars and gadgetry, as I mentioned above, funds and accelerates the development of useful products. However, a fetish for expensive wines seems to me entirely about self-placebbing or status

* I think women are let off rather lightly for this level of extravagance. If men spent two trillion dollars a year on something totally irrational – building model train sets, say – they would be excoriated for it.

† Note that pornographic films do not need a five-figure clothing budget in order to import in-season catwalk dresses without which men find it impossible to be aroused. And no man ever became aroused at the sight of a £2,000 bag.

‡ Thanks to the comedian – and astute evolutionary psychologist – Sara Pascoe for this observation.

§ Men arguably achieve the same effect by the self-administration of four pints of strong lager.

seeking, and little to do with enjoyment – after all, is a great wine really all that much nicer than a good one?¶

The Netflix documentary *Sour Grapes* is a fascinating insight into this world. A crooked, though brilliant, Indonesian wine connoisseur called Rudy Kurniawan was able to replicate great burgundies by mixing cheaper wines together, before faking the corks and the labels. He was rumbled only when he attempted to fake wines from vintages that did not exist. I am told that it is possible to detect a forged Kurniawan wine by analysing the labels, but not by tasting the wine.

I hate to say this, but Rudy was an alchemist. Several experts I have talked to in the high-end wine business regard their own field as essentially a placebo market; one of them admitted that he was relatively uninterested in the products he sold and would sneak off and fetch a beer at premium tastings of burgundies costing thousands of pounds a bottle. Another described himself as 'the eunuch in the whorehouse' – someone who was valuable because he was immune to the charms of the product he promoted.**

¶ Like L'Oréal, the slogan for the Romanée-Conti vineyard, if they were to devise such a vulgar thing, would be 'Because I'm worth it.'

** Our current obsession with wine is probably overblown. It now seems mandatory for anyone who fancies themselves to be a connoisseur to pretend to care hugely about tiny details of terroir and climate, which was not always the case. Julia Child was once asked 'What's your favourite wine?' and replied, 'Gin'.

4.8 WHAT MAKES AN EFFECTIVE PLACEBO?

One of Nicholas Humphrey's rules about what makes an effective placebo is that there must be some effort, scarcity or expense involved. Folk remedies may be effective placebos simply because the plants needed to make them aren't all that common. If there is one area that is worthy of future scientific research, this is it. At the moment, we spend many billions of dollars each year trying to improve drugs but almost nothing, as far as I can see, is being spent on the better understanding of placebos – they look too much like alchemy. I would also like to find out why the health outcomes for hospital patients appear to be better when they have a bed with a view of some trees. And what stories, I wonder, could doctors tell their patients which would maximise the power of a placebo in a consultation?

I wrote much of this book's previous chapter while suffering from an appalling bout of flu.* To reduce the symptoms in the evenings, I took a tincture known as Night Nurse,† a product with a fascinating genesis, and a textbook example of a product which doesn't

* I hope it doesn't show.

† North American readers: Night Nurse would be shelved next to NyQuil.

define the parameters of success too narrowly, so leaving room for creativity. The scientists who devised it had been briefed to come up with an effective cold and flu medicine: they succeeded, but with one problematic side effect – the formulation led to severe drowsiness. Despairingly, they were at the point of starting again when an alchemist from the marketing department came up with an idea. 'If we position the product as a *night-time* cold and flu remedy, the drowsiness isn't a problem – it's a selling point. It will not only minimise your cold and flu symptoms, but it will help you sleep through them too.' Night Nurse was born: a masterclass in the magic of reframing.

Since my wife was going to be away for a few days last month, she anxiously read me the instructions on the bottle of Night Nurse, knowing full well that I wouldn't do so myself.[‡] 'It says here that you should under no circumstances take this medicine for more than four consecutive nights,' she said nervously. Immediately I felt the placebo effect doubling its power. The fact that it should not be taken in large amounts is proof of its potency.[§] Which brings me, again, to Red Bull.

‡ More likely, I would simply take double the recommended dose 'to make sure'.

§ Another important facet of the effect, I suspect, is that it's important for any consumable product claiming to have medical powers to taste slightly weird. Things you apply to the skin would be more effective if they were to tingle or sting. A friend has told me that, in the production of the tonic wine Sanatogen, the last ingredient to be added in production was a chemical whose sole purpose was to taste slightly foul. Similarly, Diet Coke has to taste slightly more bitter than ordinary Coke, otherwise people will not believe it is a diet drink.

4.9 THE RED BULL PLACEBO

Red Bull is among the most successful commercial placebos ever produced – its powers at hacking the unconscious are so great that it is repeatedly studied by psychologists and behavioural economists all over the world, including the great Pierre Chandon at INSEAD, one of the top business schools in Europe. So potent are the drink's associations that the very presence of the logo seems to change behaviour. However, no command economy could ever have produced Red Bull, and nor could a bureaucratic large multinational – it took an entrepreneur.

The most plausible explanation for the incredible success of Red Bull lies in a kind of placebo effect. After all, it shares many of the features of a great placebo: it's expensive, it tastes weird and it comes in a 'restricted dose'. In its early days, Red Bull also benefited from repeated rumours that its active ingredient, taurine, was about to be banned. In addition to the price and the taste, the small can is particularly potent. You might expect a new soft drink to be distributed in a standard, Coke-sized can; perhaps we notice the small tin in which Red Bull is sold and unconsciously infer, 'That must be really potent stuff: they have to sell it in a small can because if you drank the full 330ml, you'd probably go doolally.'

A 2017 article in the *Atlantic* by Veronique Greenwood suggested that the risk-taking behaviour associated with cocktails containing caffeine and alcohol might be caused less by the drink itself and more by our perception of it. Greenwood explained that in 2010, the sale of such pre-packaged drinks was already banned by the FDA because of a concern that caffeine would mask the alcohol's effects. This theory seemed to be confirmed by a 2013 study which found that people consuming such drinks were twice as likely to be involved in an alcohol-linked car accident or a sexual assault as those who had drunk alcohol without caffeine.

A more recent study described by Greenwood suggests that the effect might be psychological rather than chemical. Researchers from INSEAD and the University of Michigan asked 154 young Parisian men whether they believed that energy drinks intensified the effects of alcohol. Each then drank the same cocktail of vodka, fruit juice and Red Bull, but it was labelled as either 'vodka cocktail', 'fruit juice cocktail' or 'vodka Red Bull cocktail.'

All the men were then set three tasks. First, they played a gambling game in which they won money each time they inflated a balloon a little further, with a chance that they would lose everything if it popped. The next task involved looking at photos of women and considering whether they thought they would get their numbers if they approached them in a bar. Finally, they completed surveys describing how drunk they felt and how long they would wait before driving. The results showed a clear trend: although everyone had drunk exactly the same drink, the 'vodka Red Bull group' reported feeling much drunker, took more risks than the others and were more confident when it came to approaching women. Furthermore, the effect appeared to be stronger in men who believed that mixing energy drinks and alcohol makes you take risks and reduces inhibitions, suggesting that the altered behaviour is not caused by the drink's composition but by what you *believe* it does to you. The one area where this group was more risk-averse was in relation to driving – once again, an attitude based on the perceived and not the actual effect of the drink.

According to Pierre Chandon, the branding of Red Bull, through slogans like 'Red Bull gives you wings' or the extreme sport competitions they sponsor, may not merely determine whether people buy the product but also how they respond to its name in a cocktail and how they interpret its effects. Are there lessons here that pharmaceutical companies could learn from? For instance, rather than merely putting medications in containers with a child-proof cap, might they insist that they are kept in a metal container with a combination lock? After all, even if the contents aren't particularly poisonous or potent, our inner monkey can deduce that they are – remember that the prefrontal cortex isn't involved in this decision at all, and that it is the monkey alone who decides whether a placebo works.

To those five giant industries that exist by selling mood-altering substances – alcohol, coffee, tea, tobacco and entertainment – should we add the placebo industry? After all, it's not just the purchase of cosmetics that can be explained in this way – I would contend that a large proportion of consumerism is designed to achieve the same thing. In fact, much luxury goods expenditure can *only* be explained in this way – either people are seeking to impress each other, or they are seeking to impress themselves.* Is almost everything a mood-altering substance?

* You don't have to take my word for it – I'm just some ad guy. But if you read books such as *The Mating Mind* (2000) or *Spent* (2009) by Geoffrey Miller, or *The Darwin Economy* (2001) or *Luxury Fever* (2000) by Robert H. Frank (both authors are brilliant eminent evolutionary psychologists) you will see that both come to more or less the same conclusions. Gad Saad is another very good commentator on this phenomenon, especially in *The Consuming Instinct*.

4.10 WHY HACKING OFTEN INVOLVES THINGS THAT DON'T QUITE MAKE SENSE

So signalling to ourselves or others – whether to obtain a health benefit (boosting the immune system), applying make-up (boosting confidence) or buying luxury goods (boosting status) – always seems to come accompanied by behaviours that don't make sense when viewed from a logical perspective. However, rather than being a coincidence or a regrettable by-product, it may be a necessity.

For something to be effective as a self-administered drug, it has to involve an element of illogicality, waste, unpleasantness, effort or costliness. Things which involve a degree of sacrifice seem to have a heightened effect on the unconscious, precisely because they do not make logical sense. After all, eating tasty, nutritious food is unlikely to signal anything to the immune system, since it doesn't feel out of the ordinary; drinking something foul, on the other hand, carries a greater amount of significance, since it is something you only do under unusual circumstances.

Think back to earlier in the book. Our body is calibrated not to notice the taste of pure water, because it is valuable in evolutionary terms that we notice any deviation from its normal taste. The qualities we notice, and the things which often affect us most, are the things that make no sense – at some level, perhaps it is necessary to deviate from standard rationality and do something apparently illogical to attract the attention of the subconscious and *create meaning*. Cathedrals are an over-elaborate way of keeping rain off your head. Opera is an inefficient way of telling a story. Even politeness is effectively a mode of interaction that involves an amount of unnecessary effort. And advertising is a hugely expensive way of conveying that you are trustworthy.

My contention is that placebos need to be slightly absurd to work.* All three elements that seem to make Red Bull such a potent mental hack† make no sense from a logical point of view. People want cheap, abundant and nice-tasting drinks, surely? And yet the success of Red Bull proves that they don't. Something about these three illogicalities may well be essential to its unconscious appeal, or to its potency as a placebo. If we are to subconsciously believe that a drink has medicinal or psychotropic powers, perhaps it can't taste conventionally nice. Imagine a doctor saying, 'I have some pills here to treat your extremely aggressive cancer – take as many as you like. Now, would you like them in strawberry or blackcurrant flavour?' Somehow, that last sentence doesn't quite work.

If you look at behaviours to hack the unconscious, they all seem to have an element that is wasteful, unpleasant or downright silly. Cosmetics are insanely overpriced and time-consuming to apply. Alcohol, when you think about it, doesn't *really* taste very nice: on a really hot day when you are parched, which would you honestly prefer – a glass of Château d'Yquem or a raspberry Slush Puppie? Placebo treatments like homeopathy involve a large amount of ritual and nonsense. Medicine tastes horrid.

At some point, we have to ask a vital question: do these various things work despite the fact that they are illogical, or do they work precisely *because* they are? And if our unconscious instincts are programmed to respond to and to generate behaviours precisely because they deviate from economic optimality, what might be the

* Put another way, if Richard Dawkins thinks it's a good idea, it isn't a good placebo.

† High price, small container, weird taste.

evolutionary reason for this? It seems rather like the lesson that is taught to aspiring journalists: 'Dog Bites Man' is not news, but 'Man Bites Dog' is. Meaning is disproportionately conveyed by things that are unexpected or illogical, while narrowly logical things convey no information at all. And this brings us full circle, to the explanation of costly signalling.

PART 5: SATISFICING

5.1 WHY IT'S BETTER TO BE VAGUELY RIGHT THAN PRECISELY WRONG

The modern education system spends most of its time teaching us how to make decisions under conditions of perfect certainty. However, as soon as we leave school or university, the vast majority of decisions we all have to take are not of that kind at all. Most of the decisions we face have something missing – a vital fact or statistic that is unavailable, or else unknowable at the time we make the decision. The types of intelligence prized by education and by evolution seem to be very different. Moreover, the kind of skill that we tend to prize in many academic settings is precisely the kind that is easiest to automate. Remember, your GPS is computationally much more capable than you are.

Here's a typical school maths question:

Two buses leave the same bus station at noon. One travels west at a constant 30mph, while the other travels north at a constant 40mph. At what time will the buses be 100 miles apart?*

Here's a typical real-life problem:

---

* The answer to this question in theoretical maths world is 2pm. However, in reality one of the buses was delayed by a puncture, and the other one got stuck in traffic.

I need to catch a plane leaving Gatwick at 8am. Should I get there by train, taxi or car? And at what time should I leave the house?

The odd thing about the human mind is that many people would find the first question difficult, and the second easy, yet the second question is computationally far more complex. This says more about the evolution of our brains than the difficulty of the problem. The reason is that the first question is tailor-made for computation, being what you might call a 'narrow context' problem. It assumes an artificially simplified, regularised world (where buses miraculously travel at a constant speed), it involves very few variables (all of which are numerically expressible, and allow for no ambiguity) and it has one single, incontrovertibly correct answer.

The question of how to get to Gatwick is what you might call a 'wide context' problem. It allows for vagueness and multiple right answers, and it doesn't demand absolute adherence to any precise rules. There is no formula for the solution, it allows scope for all kinds of possible 'right-ish' answers and all kinds of information can be taken into account when coming up with an answer.

These are the problems we seem instinctively better equipped to solve, but which computers find hard. If I were to delve into my unconscious and uncover some of the variables at play in my brain when I next have to get to the airport, they might include 'Is it raining?', 'How much luggage do I have?', 'How long am I going to be away for?', 'What is the *average* time via the M25 versus taking the A25?', 'What is the *variance* of journey time on the M25 versus the A25?'† and 'Does my flight leave from the North or South Terminal?'

If you think of getting to Gatwick as a narrow problem in the way your GPS does – a simple question of getting to the airport as quickly as possible – some of these factors may seem irrelevant, but they are all important in real life. The weather affects the

† The 'second-order' question which tends to get overlooked in simple optimisation models.

traffic. If I am going away for two weeks rather than one night, it affects the cost of parking, and therefore the relative cost of going by train, car or taxi – and the amount of luggage I have. The variance of travel time on the M25 matters to whether it's worth risking. And heavy luggage makes the train less appealing, especially if you are flying from the North Terminal, which is much further away from the rail station.

It's interesting that we find solving complex problems like this so easy – it suggests that our brains have evolved to answer 'wide context' problems because most problems we faced as we developed were of this type. Blurry 'pretty good' decision-making has simply proven more useful than precise logic. Now, I accept that the need to solve 'narrow context' problems is much greater today than it was a million years ago, and there's no denying the contribution that rational approaches have made to our lives – in fields such as engineering, physics and chemistry. But I would also contend that our environment has not changed all that much: most big human problems, and the majority of business decisions, are still 'wide context' problems.

The problems occur when people try to solve 'wide' problems using 'narrow' thinking. Keynes once said, 'It is better to be vaguely right than precisely wrong', and evolution seems to be on his side. The risk with the growing use of cheap computational power is that it encourages us to take a simple, mathematically expressible part of a complicated question, solve it to a high degree of mathematical precision, and assume we have solved the whole problem. So my GPS answers a narrow question like 'How long will it take to drive to Gatwick?' brilliantly, but the wider questions like 'How should I get there and when should I set off?' still remain. The GPS device has provided a brilliant answer to the wrong question. In the same way, companies can, for instance, optimise their digital advertising spend to a huge degree of precision, but this does not necessarily mean that they have answered the wider marketing questions, such as 'Why should people trust me enough to buy what I sell?'

We fetishise precise numerical answers because they make us look scientific – and we crave the illusion of certainty. But the real genius of humanity lies in being *vaguely* right – the reason that we do not follow the assumptions of economists about what is rational behaviour is not necessarily because we are stupid. It may be because part of our brain has evolved to ignore the map, or to replace the initial question with another one – not so much to find a right answer as to avoid a disastrously wrong one.

The unconscious question is not the one we are supposed to ask, and the one that might be ignored when we try to construct rational rules for decision-making. Take the case of an advertising agency pitching to potential clients. It's common for clients at such pitches to be furnished with an 'evaluation matrix' or 'scorecard', designed to ensure the process is transparent and objective. The intention is that each agency's presentation is marked on criteria such as quality of strategy, creative work, cultural fit, knowledge of sector and cost-competitiveness. The idea is that these things are scored independently and then totalled to determine the winning agency but if you ask anybody who's been involved in this procedure, they will often admit that they simply decided which agency they wanted to win and back-filled all the numbers accordingly. It may be that in situations like this what people do is substitute a completely different question to the ones they have been given, and answer that instead. This practice of answering an easier proxy question is what leads to a lot of 'irrational' human behaviour. It may fall short of being perfectly rational and it may not even be conscious, but that's not to say that it isn't clever.

Never call a behaviour irrational until you really know what the person is trying to do. A few years after university, I was living in London with a group of friends and we had each accumulated just about enough money to buy our first ever second-hand cars. We all did exactly the same thing without knowing why – we returned to the small towns where we had grown up and bought a car from someone vaguely known to our parents, whether an acquaintance, friend or relative. To an alien onlooker, this behaviour would seem bizarre, especially since second-hand cars in the provinces are slightly more expensive than they are in London. But the question we were unconsciously asking was not, 'What car should I buy, and where?', but 'Who could I find trustworthy enough to sell me a really cheap car?' We weren't trying to buy the best car in the world – we were trying to avoid the risk of buying a terrible car.

The question we were asking, 'Who can I find who won't rip me off?', was a sensible one: the one thing you can't afford to do when your budget is so tight, is fall victim to a con artist, which is why we needed someone with reputation at stake. That substitute question – 'Who do I trust to sell me an X?' – seems a perfectly reasonable proxy for buying good products. Find someone who has reputational skin in the game, ask their advice, and buy from them.‡

Like bees with flowers, we are drawn to reliable signals of honest intent, and we choose to do business where those signals are found. This explains why we generally buy televisions from shops rather than from strangers on the street – the shop has invested in stock, it has a stable location and it is vulnerable to reputational damage. We do this instinctively; what we are prepared to pay for something is affected not only by the item itself but by the trustworthiness and reputation of the person selling it.

Try this simple thought experiment. Imagine you have turned up at somebody's home to look at a second-hand car. You assess the condition of the car, which is parked on the street, and having decided it is worth £4,000, you ring on the doorbell, prepared to offer that amount. In scenario A, the door is opened by a vicar.§ In

‡ This is another reason why many theoretical models of cooperation such as the Prisoner's Dilemma are so stupid. In the real world we can choose whom we do business with. Would you happily buy a car from a vagrant you met in an alley? Obviously not.

§ You don't have to believe in God, by the way – you just have to believe that he does.

scenario B, the door is opened by a man naked except for a pair of underpants. The car has not changed, but what *has* changed is its provenance. The vicar is likely to be a man with a great deal invested in his reputation for probity, while the second man is clearly immune to any sense of shame. Would you seriously say that the amount you are prepared to pay would not increase in the first instance and decrease in the second?

It seems silly to regard this behaviour as irrational when it is really rather clever. My late mother knew absolutely nothing about cars, but had an eagle eye for people.¶ It would have been interesting to set her the task of buying ten cars based on her instincts about the people selling them, while at the same time tasking ten automotive engineers with acquiring ten cars at auction. I'm confident the cars my mother bought would have been every bit as reliable as the cars chosen by the engineers, perhaps more so.

¶ Certainly her talent for spotting a con artist would far exceed her ability to spot a doctored car.

5.2 (I CAN'T GET NO) SATISFICING

In the 1950s, the economist and political scientist Herbert Simon coined the term 'satisficing', combining as it does the words 'satisfy' and 'suffice'. It is often used in contrast with the word 'maximising', which is an approach to problem-solving where you obtain, or pretend to obtain, a single optimally right answer to a particular question.

As Wikipedia* helpfully explains, 'Simon used satisficing to explain the behaviour of decision makers under circumstances in which an optimal solution cannot be determined. He maintained that many natural problems are characterised by computational intractability or a lack of information, both of which preclude the use of mathematical optimisation procedures. Consequently, as he observed in his speech on winning the 1978 Nobel Prize, "decision makers can satisfice either by finding optimum solutions for a simplified world, or by finding satisfactory solutions for a more realistic world. Neither approach, in general, dominates the other,

* I know it is professional suicide to acknowledge Wikipedia, but in a chapter on satisficing it seems strangely appropriate. Wikipedia isn't perfect, but it is really, really good.

and both have continued to co-exist in the world of management science."' Since then, I aver, the balance has shifted. The former approach – creating a simplified model of the world and applying a logical approach – is in danger of overpowering the other, more nuanced approach, sometimes with potentially dangerous consequences: the 2008 financial crisis arose after people placed unquestioning faith in mathematically neat models of an artificially simple reality.

Big data carries with it the promise of certainty, but in truth it usually provides a huge amount of information about a narrow field of knowledge. Supermarkets may know every single item that their customers buys from them, but they don't know what these people are buying elsewhere. And, perhaps more importantly, they don't know *why* these people are buying these things from them.

A company pursuing only profit but not considering the impact of its profit-seeking upon customer satisfaction, trust or long-term resilience, could do very well in the short term, but its long-term future may be rather perilous.† To take a trivial example, if we all bought cars using only acceleration and fuel economy as a measure, we probably wouldn't do badly for the first few years, but over time, car manufacturers would take advantage of the system, producing ugly, unsafe, uncomfortable and unreliable vehicles that did fabulously on those two quantified dimensions.

There is a parallel in the behaviour of bees, which do not make the most of the system they have evolved to collect nectar and pollen. Although they have an efficient way of communicating about the direction of reliable food sources, the waggle dance, a significant proportion of the hive seems to ignore it altogether and journeys off at random. In the short term, the hive would be better off if all bees slavishly followed the waggle dance, and for a time this random behaviour baffled scientists, who wondered why 20 million years of bee evolution had not enforced a greater level of behavioural

† Modern public companies have a worryingly short lifespan as a result.

compliance. However, what they discovered was fascinating: without these rogue bees, the hive would get stuck in what complexity theorists call 'a local maximum'; they would be so efficient at collecting food from known sources that, once these existing sources of food dried up, they wouldn't know where to go next and the hive would starve to death. So the rogue bees are, in a sense, the hive's research and development function, and their inefficiency pays off handsomely when they discover a fresh source of food. It is precisely because they do not concentrate exclusively on short-term efficiency that bees have survived so many million years.

If you optimise something in one direction, you may be creating a weakness somewhere else. Intriguingly this very approach is now being considered in the treatment of cancers. I recently spoke to someone working at the cutting edge of cancer treatment. Cancer cells mutate, and therefore evolve, quickly. Trying to kill them with a single poison tends to create new mutations which are highly resistant it. The solution being developed is to target cancer cells with a chemical that causes them to be to develop immunity to it, at the expense of their immunity to other things; at that point you hit them with a different chemical, designed to attack the Achilles heel that you have created, wiping them out second rather than first time around. There is a lesson here.‡

In any complex system, an overemphasis on the importance of some metrics will lead to weaknesses developing in other overlooked ones. I prefer Simon's second type of satisficing; it's surely better to find satisfactory solutions for a realistic world, than perfect solutions for an unrealistic one. It is all too easy, however, to portray satisficing as 'irrational'. But just because it's irrational, it doesn't mean it isn't right.

‡ Exam question: is the shareholder value movement destroying capitalism?

5.3 WE BUY BRANDS TO SATISFICE

Joel Raphaelson and his wife Marikay worked as copywriters for David Ogilvy in the 1960s. We recently ate dinner at Gibson's Steakhouse at the Doubletree Hotel near O'Hare Airport in Chicago,* and talked about Joel's 50-year-old theory concerning brand preference. The idea, most simply expressed, is this: 'People do not choose Brand A over Brand B because they think Brand A is better, but because they are more certain that it is good.'† This insight is vitally important, but equally important is the realisation that we do not do it consciously. When making a decision, we assume that we must be weighting and scoring various attributes, but we think that only because this is the kind of calculation

* It is a useful heuristic to avoid airport hotels, as they have a bit of a captive market. However, there is an exception to every heuristic and the couple, as is typical of them, had chosen what must be the world's only great restaurant found at an airport hotel. It was magnificent.

† This discussion was similar to – and almost contemporary with – the work of Daniel Ellsberg in formulating the Ellsberg Paradox.

that the conscious brain understands. Although it suits the argumentative hypothesis to believe something is 'the best', our real behaviour shows relatively few signs of our operating in this way.

Someone choosing Brand A over Brand B would say that they thought Brand A is 'better', even if really they meant something quite different. They may unconsciously be deciding that they prefer Brand A because the odds of its being disastrously bad are only 1 per cent, whereas the risk with Brand B might be 2.8 per cent. This distinction matters a great deal, and it is borne out in many fields of decision science. We will pay a disproportionately high premium for the elimination of a small degree of uncertainty – why this matters so much is that it finally explains the brand premium that consumers pay. While a brand name is rarely a reliable guarantee that a product is the best you can buy, it is generally a reliable indicator that the product is not terrible. As explained earlier, someone with a great deal of upfront reputational investment in their name has far more to lose from selling a dud product than someone you've never heard of, so, as a guarantee of non-crapness, a brand works. This is essentially a heuristic – a rule of thumb. The more reputational capital a seller stands to lose, the more confident I am in their quality control. When people snarkily criticise brand preference with the phrase, 'you're just paying for the name', it seems perfectly reasonable to reply, 'Yes, and what's wrong with that?'

Imagine you're looking at two televisions. Both seem to be equal in size, picture quality and functionality. One is manufactured by Samsung, while the other is manufactured by a brand you've never heard of – let's call it Wangwei – and costs £200 less. Ideally you would like to buy the best television you can, but avoiding buying a television that turns out to be terrible is more important. It is for the second quality and not the first that Samsung earns its £200, and you are absolutely right to pay for the name in this case.

By contrast to a known brand, Wangwei has very little to lose from selling a bad television. They can't command a price premium for their name, and so their name is worthless. If a manufacturing error had caused them to produce 20,000 dud televisions, the best strategy would be to offload them on unsuspecting buyers. However, had Samsung produced 20,000 sub-par sets, they would be faced with a much greater dilemma: the reputational damage from selling the bad televisions would spill over and damage the sales of every product carrying the Samsung name, which would cost them significantly more than they would gain from the sales. Samsung would be faced with two choices: either destroy the televisions or sell them on to someone else who was less reputationally committed. It might even sell them to Wangwei, though never with its own name attached. So what's wrong with paying for that name?

The primary reason why we have evolved to satisfice in our particularly human way is because we are making decisions in a world of uncertainty, and the rules for making decisions in such times are completely different from those when you have complete and perfect information. If you need to calculate the hypotenuse of a triangle and you know one interior angle and the length of the other two sides, you can be perfectly correct, and many problems in mathematics, engineering, physics and chemistry can achieve this level of certainty. However, this is not appropriate to most of the decisions we have to make. Such questions as whom to marry, where to live, where to work, whether to buy a Toyota or a Jaguar or what to wear to a conference don't submit to any mathematical solution. There are too many future unknowns and too many variables, many of which are either not mathematically expressible or measurable. Another good example of a decision that has to be taken subjectively is whether we choose to buy an economical or a high-performance car. In general there is a trade-off between these two attributes. Do you sacrifice economy for performance or performance for economy?‡

‡ Your decision will also greatly depend on whether you already own a car. Few people, I think it's fair to say, own a Bugatti Veyron as their everyday runabout. Around 80 per cent of Rolls-Royce owners also own a Mercedes. Remember the story about recruiting people in groups? The greater the number

Imagine you're living in the wild. You notice some extremely attractive cherries high up in a tree, but you know that, delicious and nutritious as they would be, there is a small risk that in attempting to pick them you will fall to your death. Let's say the risk is one in 1,000, or 0.1 per cent. A crude mathematical model would suggest that this risk only reduces the utility of the cherries by a tenth of a per cent,[§] but this would be a foolish model to use in real life – if we routinely exposed ourselves to risks of this kind, we would be dead within a year. You'd only take this risk if you were *very* hungry – if there were a correspondingly high risk that you would die of starvation if you were *not* to eat the cherries, climbing the tree might make sense. However, if you weren't starving and you knew that perfectly nourishing, if less tasty, foods were available elsewhere at a lower risk of fatality, you'd wander off and find a safer source of nourishment.[¶] Remember, making decisions under uncertainty is like travelling to Gatwick Airport: you have to consider two things – not only the expected average outcome, but also the worst-case scenario. It is no good judging things on their average expectation without considering the possible level of variance.

---

of cars you own, the greater the variance of choice. If someone owns three Corvettes, they really need to buy a small electric car.

§ The net value of the cherries minus the small chance that you'll never survive to eat them.

¶ The plotline to almost every successful work of literature and every interesting film – whether a rom-com or an action movie – involves exceptional, high-adrenaline moments when characters are forced to throw caution to the wind. To create these moments in an age of technology, Hollywood is forced to fall back on rather hackneyed devices – for example, one of the protagonists will hold up a mobile phone and exclaim, 'Damn, no signal.' This is to prevent the audience losing patience with the plot by simply thinking, 'I don't get it – why don't you just call the police?'

Evidence that similar mental mechanisms also apply to human purchasing decisions can be found by looking at the data on eBay. In a simplistically logical world, a seller with an approval rating of 95 per cent** would nonetheless be able to sell goods perfectly successfully if they were 10 per cent cheaper than goods offered by people with 100 per cent approval ratings. However, a quick glance at the data shows this is not the case. People with approval ratings below 97 per cent can barely sell equivalent goods for *half* the price of sellers with a track record of 100 per cent satisfaction.

Logically you might think we should accept a 5 per cent risk of our goods not arriving in exchange for a 15 per cent reduction in cost, but the lesson proved by these statistics is that we don't: once the possibility moves beyond a certain threshold, we seem unable to take the risk at any price. If Amazon were to try to operate in a country where 10 per cent of all posted goods were stolen or went missing, virtually no discount would be high enough for them to sell anything at all.

This example illustrates that, when we make decisions, we look not only for the expected average outcome – we also seek to minimise the possible variance, which makes sense in an uncertain world. In some ways, this explains why McDonald's is still the most popular restaurant in the world. The average quality might be low, compared to a Michelin-listed restaurant, but so is the level of variance – we know exactly what we're going to get, and we always get it. No one would say that a meal they had had at McDonald's was among the most spectacular culinary experiences of their lives, but you're never disappointed, you're never overcharged and you never get ill. A Michelin three-star restaurant might provide an experience that you will cherish for the rest of your life, but the risk of disappointment, and indeed illness, is also much higher.††

** Quite low, by eBay standards.

†† In 2011 one of Britain's finest restaurants, a three-star Michelin establishment, suffered the worst norovirus outbreak ever experienced by a single restaurant.

In a world of perfect information and infinite calculating power, it might be slightly suboptimal to use these heuristics, or rules of thumb, to make decisions, but in the real world, where we have limited trustworthy data, time and calculating power, the heuristic approach is better than any other alternative.

For instance, a cricketer catching a high-flying ball does not calculate its trajectory using quadratic equations, but instead uses a rule of thumb known as the 'angle of gaze' heuristic, looking upwards at the ball and moving towards it in such a way that the upward angle of their gaze remains constant. In this way, though he may pay the price of moving in a slight arc rather than in a straight line, he will hopefully position himself at the point on the ground where the ball is likely to land. There are several reasons why we use a heuristic of this kind. A fielder would, of course, have no time to perform mathematical calculations, even if a calculator were available, and moreover, even if he had enough time and calculating power, he simply wouldn't have enough data without knowing the velocity or the angle at which the ball was hit to calculate its trajectory to any level of accuracy. The batsman who hit the ball probably wouldn't know, either.‡‡

Rather oddly, this did not seem to imperil its ranking. Presumably so long as the *jus* and *tapenades* are all impeccably handmade, the Michelin inspectors do not consider three days spent shitting out your innards as significantly detracting from the dining experience.

‡‡ And, even if he did know, he wouldn't tell you.

5.4 HE'S NOT STUPID, HE'S SATISFICING

On 15 January 2009, in an incident now known as the 'Miracle on the Hudson', Captain Chesley Sullenberger demonstrated the value of heuristics when, after his aircraft had both its engines disabled by a bird strike, he reacted quickly and safely landed on the Hudson River. It is possible to listen to Sullenberger's conversations with air traffic control on YouTube: between attempts to restart the engines, he communicates with the departure airport. Having immediately rejected the possibility of returning to LaGuardia, correctly as it turns out, he is offered the possibility of landing at Teterboro Airport, which is in New Jersey over to his starboard side. In a little more than 20 seconds he decides that this option is also impossible – again, a decision made with a heuristic rather than a calculation. He did not retrieve a scientific calculator from his briefcase, input the flight speed, altitude and rate of descent and then calculate the likely distance to runway one at Teterboro, but instead did something far quicker, easier and more reliable.

A former US Air Force fighter pilot, Sullenberger was a glider pilot in his spare time, and all glider pilots learn a simple instinctive rule which enables them to tell whether a possible landing site on the ground is within their reach. They simply place the glider in the shallowest possible rate of descent and look through the windshield: any place which appears to be moving downwards in the field of view is somewhere they can safely land, while anywhere on the ground that appears to be moving upwards is too far away. It was by deploying this rule that he was able to decide within seconds that the Hudson River was the only feasible landing site.

In the event his decision could not have been bettered. There were no fatalities, and only a few minor injuries. It's true that, had he successfully landed at Teterboro, he might have saved the aircraft, but had he tried and failed to land there, it is doubtful that anyone could have survived.

It isn't always clear which heuristic rules are learned and which are innate, but everyday life would be impossible without them. A truck driver reversing an articulated lorry into a narrow driveway achieves what seems like a spectacular feat of judgement through the use of heuristics, not by calculation. We drive our cars heuristically, we choose our houses heuristically – and we probably also choose our partners heuristically.* Even when a solution *might* be calculable, heuristics are easy, quick and well-aligned with our perceptual equipment, and in the majority of occasions where the right solution is incalculable, they are all we've got.

Heuristics look second-best to people who think all decisions should be optimal. In a world where satisficing is necessary, they are often not only the easiest option but the best.

* Smell almost certainly plays a much greater role in attraction than we are aware of. One experiment suggests we are attracted to the smell of people who have an immune system complementary to our own.

5.5 SATISFICING: LESSONS FROM SPORT

I have always been intrigued by the scoring systems in different sports, and by the degree to which they contribute to the enjoyment of any game. As a friend of mine once remarked, had tennis been given the same scoring system as basketball it would be tedious to play and even worse to watch: if you glanced at your TV and saw Djokovic leading Murray 'by 57 points to 31', you would shrug and change channels to something more exciting.*

Tennis scoring isn't quite socialist – one player can demolish another – but, in such uneven cases, the contest is over in a mercifully short time. There is, however, a kind of social security system in the sport's scoring system, which means that for the duration of any match, the losing player feels he might still be in with a chance. It's frankly genius.

The system of watertight games and sets means that there is no difference between winning a game to love or after several deuces. A 6–0 set counts as a set, just as a 7–5 win does. This means that the losing player is never faced with an insurmountable mountain to climb. The scoring system also ensures variation in how much is at stake throughout the game; someone serving at

* An unsubtitled version of *Last Year at Marienbad*, for example.

30–0 is a relatively low-engagement moment, while a crucial break point has everyone on the edge of their seats. This varies the pitch of excitement, and consequently makes the game more enjoyable for players and spectators alike.†

Another feature in the scoring system of many compelling games is where aiming for the highest score comes with a high degree of concomitant risk. Shove ha'penny works in this way, as does bar billiards, where the highest-scoring pot sits behind an unstable black mushroom (technically a 'skittle'), which wipes your entire score if it is knocked over. This jeopardy may explain why darts is an enjoyable spectator sport, while archery isn't. In archery the scoring is concentric. You simply aim for the bullseye, which scores 10, and if you narrowly miss you get 9. Miss the 9 and you get 8, and so on. The only strategy of the game is to aim for the 10 and hope – it is a perfectly logical scoring system, but it doesn't make for great television. The dartboard, by contrast, is not remotely logical, but it's somehow brilliant. The 20-point sector sits between the dismal scores of 5 and 1.

Most darts players aim for the treble 20, because that's what the professionals do. However, for all but the best darts players this is a mistake: if you are not very good,‡ your best approach is not to aim at treble 20 at all, but instead to aim at the south-west quadrant of the board, towards treble 19 and 16. You won't get 180 that way, but nor will you score 3. It is a common mistake in darts to assume that you should simply aim for the highest possible score – you should also consider the consequences if you miss.

Many real-life decisions have a scoring rubric that is more like darts than archery. For instance, in deciding whom to marry, aiming for the best may be less important than avoiding the worst – rather than trying to maximise an outcome, you may seek a pretty good all-round solution with a low chance of disaster. A darts player

---

† Notice that, if you *don't* understand the scoring system, tennis is actually rather boring to watch.

‡ Or if you are drunk.

repeatedly aiming for the south-western quadrant of the board would look insane to many onlookers, who might say, 'You're supposed to aim for the triple 20 – it's the highest score on the board,' but an approach seeking to minimise variance or minimise downsides often involves behaviour that seems nonsensical to those who don't understand what the actor is trying to do.

By the same token, to someone who assumes that holidaymaking is a lifelong quest to find new experiences, returning annually to the same resort may seem ridiculous; it is on the other hand an extremely good approach if you want to avoid a bad holiday. Habit, which can often appear irrational, is perfectly sensible if your purpose is to avoid unpleasant surprises.

Social copying – buying products or adopting behaviours and fashions that are popular with others – is another safe behavioural approach. After all, the bestselling car in Britain is unlikely to be terrible. Another reliable risk reduction strategy when making decisions under uncertainty is simply to substitute a different question from the one that conventional logic assumes you should be asking. So you would not ask 'What car should I buy?', but 'Whom can I trust to sell me a car?' Not 'What's the best television?', but 'Who has most to lose from selling a bad television?' Or not 'What should I wear to look great?', but 'What's everyone else going to be wearing?'

A common approach when recruiting staff is to ask your existing employees to recommend people – indeed, this is how most entry-level jobs in mid-sized companies seem to be filled. This might seem to involve fishing in a very narrow pool, and it is, but a personal recommendation from an existing staff member is a good way to avoid hiring someone terrible. People are keen to do favours for their mates, of course, but no one is going to jeopardise their reputation at work by recommending an alcoholic, a kleptomaniac or an arsonist. Third-party recommendations are not perfect or remotely scientific, but they are rarely catastrophic.

Many apparent paradoxes of consumer behaviour are best explained by similar mental mechanisms. A few years ago we discovered that men were reluctant to order a cocktail in a bar – in part because they had no foreknowledge of the glass in which it would be served. If they thought there was even a slight chance that it would arrive in a hollowed out pineapple, they would order a beer instead. One remedy was to put illustrations or pictures of the drinks on the menu; some trendy venues have since solved the problem by serving all their cocktails in mason jars. The same sort of mental calculus explains why it is so difficult to get people to move their current account from one bank to another paying a higher rate of interest, or to shift their broadband provision. A 1 per cent chance of a nightmarish experience dwarfs a 99 per cent chance of a 5 per cent gain.

5.6 JFK VS EWR: WHY THE BEST IS NOT ALWAYS THE LEAST WORST

I once asked, over Twitter, whether there were any clear advantages to flying to JFK Airport in New York rather than Newark.* Other than a string of replies from New Yorkers with an inbuilt disdain for anything in New Jersey† there seemed to be few arguments for using JFK: Newark is closer to Manhattan, and risks fewer road-works or delays on the journey. Richard Thaler, one of the world's most eminent decision scientists, tweeted me with strong support for Newark.‡ If it were only informed consumers making the choice, Newark would surely be the more successful airport, yet JFK is the more common choice. Ironically it may be more popular simply because it is more popular – if that sounds like nonsense, bear with me.

Because JFK is more popular, it is seen as a less eccentric choice. Flying to JFK is the equivalent of buying an IBM mainframe in 1978:

* I had never been able to understand the popularity of JFK, and wondered if there was some elusive upside that had escaped me.

† Being British, I do not generally make hairsplitting distinctions between rebellious former colonies.

‡ Though, since he was born in New Jersey, he may have been affected by bias towards his home state.

an easy default. The great thing about making the 'default' choice is that it feels like not making a decision at all, which is what businesspeople and public sector employees tend to really like doing – because every time you don't visibly make a decision, you've ducked a bullet. Newark requires a rational justification: 'Why is my flight going to Newark; why not JFK?' By contrast, the sentence, 'I've booked you a flight to JFK', rarely meets with the question, 'Why JFK? What's wrong with Newark?'

So, imagine you are the personal assistant to a grumpy boss in London. He or she asks you to book a flight to New York. You have two choices.

1. Book the boss on a flight to JFK, hand over the tickets and relax.
2. Book them on a flight to Newark and cross your fingers.

There's a large chance that if you take option 2 – the better decision – you will come out quite well. If your grumpy boss notices the ease of the journey from Newark and the friendliness of immigration staff, they may remember to thank you on their return; you might even get 'Nice choice – remind me to use Newark next time.' However, it is unlikely that your boss would buy you a case of vintage champagne and immediately award you a four-figure bonus – a thank-you is the best you can get.

But flights are delayed or cancelled – and the reason that you would need to keep your fingers crossed after choosing option 2 is that, when things do go wrong, as they sometimes will, the difference between choosing options 1 and 2 becomes more stark. If the flight from JFK is delayed by three hours, your boss will blame the airline, but if the flight from Newark is, your boss will probably blame you, because with option 2 you made a noticeable decision – you deviated from the default. He might say, 'This wouldn't have been a problem if you had booked me from JFK – the flights were fine there. What were you thinking, booking me a flight from this weird airport, you fool?'

Blame, unlike credit, always finds a home, and no one ever got fired for booking JFK. By going with the default, you are making a worse decision overall, but also insuring yourself against a catastrophically bad personal outcome. In his book *Risk Savvy* (2014), the German psychologist Gerd Gigerenzer refers to this mental process as 'Defensive Decision-Making' – making a decision which is unconsciously designed not to maximise welfare overall but to minimise the damage to the decision maker in the event of a negative outcome. Much human behaviour that is derided as 'irrational' is actually evidence of a clever satisficing instinct – repeating a past behaviour or copying what most other people do may not be optimal, but is unlikely to be disastrous. We are all descended from people who managed to reproduce before making a fatal mistake, so it is hardly surprising that our brains are wired this way.

In institutional settings, we need to be alert to the wide divergence between what is good for the company and what is good for the individual. Ironically, the kind of incentives we put in place to encourage people to perform may lead to them to be unwilling to take any risks that have a potential personal downside – even when this would be the best approach for the company overall. For example, preferring a definite 5 per cent gain in sales to a 50 per cent chance of a 20 per cent gain. Why else do you think Management Consultancies are so rich?

PART 6: PSYCHOPHYSICS

6.1 IS OBJECTIVITY OVERRATED?

You may have never heard the term 'psychophysics', which is essentially the study of how the neurobiology of perception varies among different species, and how what we see, hear, taste and feel differs from 'objective' reality. For instance, different species, as I will soon explain, perceive colour very differently, since receptors in the eyes are sensitised to different parts of the light spectrum. More importantly, our different senses – though we don't realise this – act in concert; what we see affects what we hear, and what we feel affects what we taste.*

A few years ago, the British chocolate manufacturer Cadbury's received a large number of customer complaints, claiming that they had changed the taste of their Dairy Milk brand. They were at first baffled, because the formulation hadn't been altered for years.

* A friend of mine who lives overseas is sadly losing her hearing, though I didn't realise this at all when we met recently – she had learned to lip read astonishingly well. But what was truly fascinating was that she hadn't realised that her hearing was failing until late in the process, because she was unconsciously learning to lip read and was hearing sounds which in reality she could only 'see'.

However, what they had done was change the shapes of the blocks you would break off a bar, rounding their corners. And smoother shapes taste sweeter. Truly.

Nothing about perception is completely objective, even though we act as though it is. When we complain that a room is hot, there may be no point at which we agree about what 'hot' means; it may merely mean 'a few degrees warmer than the room I was in previously, to which I have become acclimatised'. 'Time flies when you are having fun' is an early piece of psychophysical insight. To your watch, an hour always means exactly the same thing, regardless of whether you are drinking champagne or being waterboarded. However, to the human brain, the perception of time is more elastic.†

In some businesses, psychophysics is a more valuable discipline than physics, and in many industries you need to master both. The airline industry is a good example: along with the physics of flight, you need to understand the psychophysics of taste, because food tastes very different at altitude, meaning that meals that are pleasant on the ground can be boring in the air.‡ Endless complaints about airline food, which were once a staple of stand-up comedy, may have been unfair: it wasn't that the food was bad, but it was the wrong type of food to eat at 30,000 feet.

The new Boeing 787 Dreamliner is, in many ways, a triumph of psychophysics. Lighting, pressurisation and humidity all mitigate the effects of jet lag. Moreover, visual illusions – in particular a spacious entranceway – create an impression of spaciousness; it is actually 16 inches narrower than a Boeing 777, but to many passengers it feels significantly wider. Adding a little space when people enter the aircraft creates an impression of airiness that carries through into the main cabin, even though the main cabin is no less densely packed than usual. Blake Emery, the psychologist

† I spent 24 hours in a Qatari prison once; it seemed like a month.

‡ Memo to the world's airlines: 'More curries please.' Indian food tastes magnificent at altitude.

in charge of product differentiation at Boeing, explains that his team were 'looking for things that people really couldn't articulate' that might improve the experience. No one actually knows the humidity and air pressure inside an airliner, but these things have a large effect on how people feel. Historically, airliners were designed with the airline's accountant rather than the passenger in mind – all they cared about was cost and capacity. So Boeing's attempts to differentiate the passenger experience were a bold move.§

Engineers and accountants are prone to ignore the human side of their creations, and this is not always wrong – if you are building an unmanned space rocket or a bridge, it is possible to define success in objectively defined measures. However, if you are designing anything where human perception plays a part, you need to play by a different set of rules. For instance, a bridge that supports a prespecified volume of traffic and survives in all reasonable climatic conditions can be said to be a good bridge. It would, of course, be *better* if it were an attractive bridge rather than an ugly one, but that is a secondary consideration – in designing a bridge, there is little scope for alchemy.

In other projects, like designing a train service or a tax system, or in painting the lines on roundabouts, it is impossible to define success *except* in terms of human behaviour. Here there is generally some potential for alchemy, since perception, rather then reality, is what determines success. Even giving a tax a different name can have a colossal effect on whether people are willing to pay it.¶

§ My own subjective experience, having flown several times on a 787, is that it will pay off. My last trip, to Los Angeles, was the first time I have ever crossed the Atlantic and experienced no ill effects from jet lag at all.

¶ This is why some people propose that inheritance tax, or death duties, should be renamed as a 'windfall tax' and levied on the recipients, rather than the estate of the deceased.

6.2 HOW TO BUY A TELEVISION FOR YOUR PET MONKEY

You probably aren't aware of this, but your television is cheating you. The same applies to the computer screen* and also to the colour pictures in magazines. Not everything on an LCD screen involves deceit: when the screen shows pure blue, green or red, it is more or less telling the truth – blue lights produce pure blue photons, green lights produce green photons and red lights produce red photons. Each pixel on the screen contains three LCD lights – one in each of those three colours. If the red lights alone are displayed, the screen is red.

But yellow on television is a big fat lie. It may look yellow, but it isn't really – it's a mixture of red and green light, which hacks our optical apparatus to make us think we are looking at something genuinely yellow. The yellow is created in our brains, not on the screen. Colour mixing is a biological, not a physical phenomenon – you can't mix green and red photons to make yellow photons, but by sending an image of a mixture of red and green photons to the brain in the right ratio, the resulting stimulus is indistinguishable from that of yellow photons, and you see yellow as a result. Even so, yellow isn't as much of a lie as purple† is – yellow does, at least,

* On which I hope you are not reading this book!

† Technically, magenta.

exist on the light spectrum.‡ Purple does not exist at all: indigo and violet are in a rainbow, but magenta isn't – the colour exists only in our heads.

The reason for all this is that humans – and indeed all apes – have trichromatic vision. We have three sets of cones (or colour sensors) in our retinas, each of which is sensitive to a different part of the colour spectrum; the brain then constructs the rest of the spectrum by extrapolating from the relative strength of these three. In the case of purple, which occurs when the red and blue sensors but not the green ones are triggered, the brain creates a colour to fill the gap.§ Three colours are hence enough to recreate a wide (though not quite complete) spectrum of colours – not on the screen, but inside our heads; including some colours that aren't really there at all.

Because colour mixing is a biological phenomenon, how it works depends on the species (and sometimes the individual) which

Many female marmosets see in three colours, but males can only see in two.

‡ It's the 'York' in 'Richard Of York Gave Battle In Vain' (or the 'y' in 'Roy G. Biv').

§ If your brain were more objective, rather than showing you purple, it would display a patch of flickering grey with the words 'system error' on it.

performs it. If lemurs and lorises bought televisions, it would be advisable to produce a cheaper dichromatic LCD TV for them – these primates construct their colour spectrum from green and blue alone, so you could omit the component of each pixel that generates red light.

It's a good thing that marmosets don't buy televisions, as it could be a source of marital discord. Females and males of the species have completely different colour perception – many females see in three colours, while males only see in two. A female might come home having spent £800 upgrading their 65-inch OLED to one of the new hyper-realistic tri-chromatic models, but her mate would complain that 'it looks just the same as the old one'.¶ Your best bet is to keep a pet owl monkey, because they would be perfectly happy with a black-and-white TV: like many other nocturnal mammals, they don't see colours at all.

No one advertises televisions as being 'designed for higher primates', but it would be perfectly accurate to do so. The lesson to take from this is that it is possible for something to be objectively wrong but subjectively right. TVs are designed around how we *see*, not what they *show*. There is a lot of clever engineering involved in making a television,** but the real genius in it is psychological alchemy, not technology – without an understanding of how humans perceive colour, making one would be almost impossible.

As I have argued, psycho-logic and psychophysics need to be applied not just to the design of televisions, but also to welfare programmes, tax, transportation, healthcare, market research, the pricing of products and the design of democracy. There is no point in struggling to create changes in objective reality if human per-

¶ This would be a rare instance of a male primate criticising a female for spending too much on a TV.

** The three men who invented the blue LED (Isamu Akasaki, Hiroshi Amano and Shuji Nakamura) received the 2014 Nobel Prize for Physics.

ception can't see it, so all these things need to be perception optimised for humans. Moreover, just as with the colour purple, we should remember that if you design something in a certain way, people can perceive something which doesn't exist in reality.††

What really *is* and what we *perceive* can be very different.

This is where physical laws diverges from psychological ones. And it is this very divergence which makes Alchemy possible.

†† An advertising friend of mine moved to Majorca. 'It's a wonderful location, because on the overnight ferry I can get to France in an hour and to Barcelona in an hour.' He paused. 'Well, it's actually nine hours, but I'm asleep for eight of them.'

6.3 LOST AND GAINED IN TRANSLATION: REALITY AND PERCEPTION AS TWO DIFFERENT LANGUAGES

There is a whole academic discipline devoted to the idea that human behaviour can be modelled as if it were a physical phenomenon: it's called economics. However, the simple fact is that, in all facets of human behaviour, reality and perception are like two different languages, each with concepts that are more or less untranslatable into the other.*

Emotions are stranger still and, like magenta, are produced in the mind. If a tree falls in the forest and no one hears it, does it make a sound? Yes, because mechanical sensors could still record that sound. But if a car lingers too long at a green traffic light and there is no one waiting behind to get angry, is it an annoyance? No, because annoyance is a perceptual concept that is confined to living things. Obviously, perception and reality sometimes closely parallel each other, but at other times there is a complete disconnect between the two that is similar to a language gap. If you take any two languages at random, you may find that the two are sometimes immensely different, containing concepts unique to one or the

* Just as you can't translate perceptual purple into purple photons. And you can't translate ultra-violet photons into human vision.

other.† Other language-pairs are similar,‡ perhaps confusingly so, but each situation brings its own problems, and in both cases you can experience translation errors.

The job of a designer is hence that of a translator. To play with the source material of objective reality in order to create the right perceptual and emotional outcome.

† Native American languages are like this.

‡ Spanish and Portuguese, say.

6.4 MOKUSATSU: THE A-BOMB, THE H-BOMB AND THE C-BOMB

Translation errors can be expensive, and at times gruesomely so. The following is from'Mokusatsu: one word, Two Lessons', a declassified article in the National Security Agency's *Technical Journal* (Fall 1968):

In July of 1945 allied leaders meeting in Potsdam submitted a stiffly worded declaration of surrender terms and waited anxiously for the Japanese reply. The terms had included a statement to the effect that any negative answer would invite 'prompt and utter destruction'. Truman, Churchill, Stalin, and Chiang Kai-Shek stated that they hoped that Japan would agree to surrender unconditionally and prevent devastation of the Japanese homeland and that they patiently awaited Japan's answer.

Reporters in Tokyo questioned Japanese Premier Kantaro Suzuki about his government's reaction to the Potsdam Declaration. Since no formal decision had been reached at the time, Suzuki, falling back on the politician's old standby answer to reporters, replied that he was withholding comment. He used the Japanese word 'mokusatsu', derived from the word for 'silence'. However, the word has other meanings, quite different from that intended by Suzuki.

Alas, international news agencies saw fit to tell the world that in the eyes of the Japanese government the ultimatum was 'not worthy of comment.' US officials, angered by the tone of Suzuki's statement and obviously seeing it as another typical example of the fanatical Banzai and Kamikaze spirit, decided on stern measures. Within ten days the decision was made to drop the atomic bomb, the bomb was dropped, and Hiroshima was leveled.

Depending on the context, *mokusatsu* can mean many things. It is derived from words meaning 'silence' and 'death', and can mean anything from 'I cannot say anything at this time' to 'I have nothing to say because I am nonplussed' or 'I contemptuously refuse to dignify your proposal with a response.'

Japanese is a highly context-sensitive language, but then so are all languages. In British English, when said in the right context and tone, 'You stupid fucking idiot' can be a term of affection – something that can wrongfoot Americans, who mostly speak the same language but tend to interpret it more literally.*

In translation, it is an enormous mistake to assume that what the translator conveys is what the speaker intended, and it is

* I have spoken British English all my life, but I am still not sure I could reliably explain the distinction between 'quite' meaning 'very' and 'quite' meaning 'somewhat'. It's just one of those things you grow up understanding – like the many advantages of having a proper monarchy and not allowing everyone to own guns, you either get it or you don't. I once gave a talk in New York where I used the phrase 'feel like a bit of a twat'. In British English the word 'twat' is close in meaning to (though still stronger than) 'twit', but In US English using 'twat' or 'twot' is tantamount to dropping the C-bomb. As a result my talk, which had been filmed, needed to be edited before being sent out. Afterwards, someone came up and told me they thought my talk had been 'quite good', which in British English means 'kind of okay but nothing special'. I was waiting for tips on how to improve it, before I realised that they'd liked it a lot: in American English, 'quite' is an intensifier, like 'very' or 'really'. In British English it is occasionally an intensifier ('quite excellent'), but mostly it is a modifier ('quite interesting').

equally foolish to assume that what you intended to say is what will be understood. Perhaps the most famous example of mistranslation occurred on a US presidential visit to Poland in 1977. Giving a short speech on the tarmac shortly after landing in Warsaw, President Carter was heard by his hosts to say that he 'had left America, never to return' and that 'his affection for the Polish people was so great that he wanted to have sex with them.'†

The story is often told as an illustration that the translator was some idiot and not up to the job, but this isn't so – Steven Seymour was a brilliant translator, who had earlier translated the poems of W.H. Auden into Russian and had a true connoisseur's appreciation of Polish poetry. However, his study of poetic Polish had unfortunately made him overfamiliar with the more antiquated nineteenth-century (or earlier) Polish vocabulary that modern Poles no longer used, or at least not in the same sense.‡

Russian was Seymour's first language, while Polish was only his fourth. This would not have mattered, were it not for the fact that Polish is misleadingly similar to Russian in much of its vocabulary and grammar,§ but occasionally stubbornly different in its meaning. Translators call these misleadingly similar words 'false friends', since it is very easy to assume they mean something they don't.¶

---

† Carter's comments, as read by the interpreter, were often backtranslated as 'I desire to know the Poles carnally'. This was, if anything, a bit of a euphemism.

‡ In the same way that modern-day Amish would probably no longer choose 'Intercourse' as the name for a town in Pennsylvania.

§ Even though Poles probably don't like to acknowledge this.

¶ '*Constipado*' in Spanish is a false friend if you speak English or French: it is very easy to translate as 'constipated', as one French translator famously did, forgetting that to a Spaniard it means 'bunged up with a cold'.

Languages that are similar may be prone to greater misunderstandings – in Latin American countries, for instance, Spanish words may take on different meanings: 'Your wife is a tremendous whore' would seem an odd way to thank your host after dinner, except that in *some* countries, the formal word for 'hostess' has acquired that meaning.

Strangely, one of the greatest sources of linguistic confusion arises between British and Dutch speakers of English. The Dutch are, almost universally, fluent in English:** their grasp of idiom is superb, their accents are flawless, and they have a similarly cynical sense of humour to the British. After an evening with a couple of Dutch contemporaries, you would have no consciousness of a language barrier, and would find it hard to believe that any misunderstanding could arise. However, misunderstandings are all too common, because Dutch conversation tends to be astoundingly direct, while British English is oblique and often coded to the point of derangement. In a business context a Dutchman might say, 'We tried that and it was shit, so we won't do it again,' while an Englishman intending to say the same thing might say, 'I think it might be a little while before we try that again.'

Eventually the Dutch compiled a sort of phrasebook, which translates British English into Dutch English.

WHAT THE BRITISH SAY	WHAT FOREIGNERS UNDERSTAND	WHAT THE BRITISH MEAN
I hear what you say	He accepts my point of view	I disagree and do not want to discuss it further
With the greatest respect	He is listening to me	You are an idiot

** Perhaps helped by the fact that, in the Netherlands, English-language films are subtitled rather than dubbed.

WHAT THE BRITISH SAY	WHAT FOREIGNERS UNDERSTAND	WHAT THE BRITISH MEAN
That's not bad	That's poor	That's good
That is a very brave proposal	He thinks I have courage	You are insane
Quite good	Quite good	A bit disappointing
I would suggest	Think about his idea, but I should do what I like	Do it or be prepared to justify yourself
Oh, incidentally/ by the way	That is not very important	The primary purpose of our discussion is
I was a bit disappointed that	It doesn't really matter	I am annoyed that
Very interesting	They are impressed	That is clearly nonsense
I'll bear it in mind	They will probably do it	I've forgotten it already
I'm sure it's my fault	Why do they think it was their fault?	It's your fault
You must come for dinner	I will get an invitation soon	It's not an invitation, I'm just being polite
I almost agree	He's not far from agreement	I don't agree at all
I only have a few minor comments	He has found a few typos	Please rewrite completely
Could we consider some other options?	They have not yet decided	I don't like your idea

The barrier between the English spoken by the Dutch and the British is a decent metaphor for the relationship between reality and perception – in some respects they are similar, but in other contexts they can diverge wildly. Again, this distinction – the gap between the message we intend to send and the meaning that is attached to it – matters a great deal. Often, we are baffled by people's behaviour. 'I told him this, and he did that.' We think they are being irrational, but the reality is that they didn't hear what we think we said.

In the same way, you cannot describe someone's behaviour based on what *you* see, or what *you think they see*, because what determines their behaviour is what *they* think they are seeing. This distinction applies to almost anything: what determines the behaviour of physical objects is the thing itself, but what determines the behaviour of living creatures is their perception of the thing itself.

The reason this matters so much is that most models of human behaviour and most economic models are blind to this distinction. It won't surprise you to know that I am sceptical about the promise of 'big data', which is frequently promoted as though it were some kind of panacea. Like many things that emerge from the technology sector, we become so drunk on the early possible benefit of a technology that we forget to calculate the second-order problems.†† The evangelists of big data imply that 'big' equals 'good', yet it by

†† We welcomed the invention of email because it gave us the power to communicate with the world instantaneously and for free, but we forgot to ask what the consequences might be if everyone else on the planet was similarly free to communicate with us.

no means follows that more data will lead to decisions that are better or more ethical and fair.‡‡

To use the analogy of the needle in the haystack, more data does increase the number of needles, but it also increases the volume of hay, as well as the frequency of false needles – things we will believe are significant when really they aren't. The risk of spurious correlations, ephemeral correlations, confounding variables or confirmation bias can lead to more dumb decisions than insightful ones, with the data giving us a confidence in these decisions that is simply not warranted.

A large tech company recently developed an AI system to sift applications for jobs, but it rapidly developed extreme gender prejudices – marking people down if their CV mentioned, say, participation in *women's* basketball. With AI, of course, you cannot always be sure of its reasoning: it may have noticed that more senior employees were men, and so taken maleness as a predictor of success.

Another company using a big data approach discovered a variable that was vastly more predictive of a good employee than any other: it wasn't their level of educational attainment or a variable

‡‡ The ethnographer Tricia Wang even suggested in her 2016 TEDxCambridge talk that the quantification bias created by big data led to the near death of Nokia as a handset manufacturer. All their data suggested that people would only spend a certain proportion of their salary on a phone handset, so the market for smartphones in the developing world would be correspondingly small. Wang noticed that, once people saw a smartphone, their readiness to spend on a handset soared. Her findings were ignored as she had 'too few data points'. However, in reality, all valuable information starts with very little data – the lookout on the *Titanic* only had one data point ... 'Iceberg ahead', but they were more important than any huge survey on iceberg frequency.

Do ice cream sales drive violent crime?

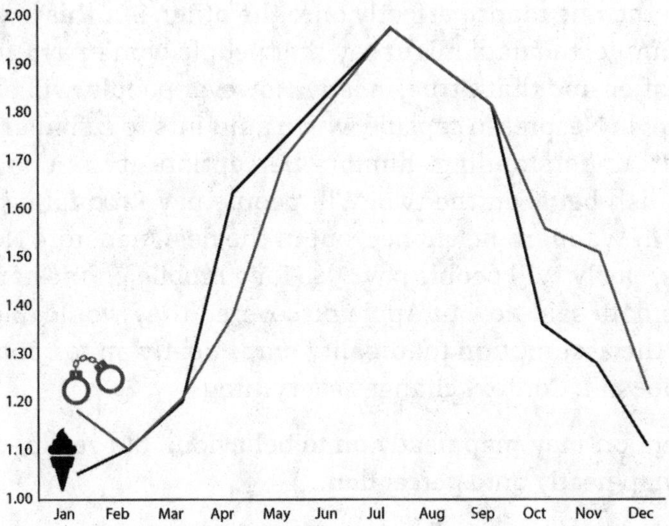

The confounding variable here, missing from the data, is the weather, which explains the spurious correlation. To a dumb alogorithm, it might appear from the graph that ice cream consumption drove people to commit crimes. However, the real reason is simple: people consume more ice cream when it is sunnier, and they also commit more crimes on warm evenings.

on a personality test – no, it turned out that the best employees had overwhelmingly made their online application using either Google Chrome or Firefox as their browser, rather than the standard one supplied on their computers. While I can see that replacing a browser on a laptop may be indicative of certain qualities – conscientiousness, technological competence and the willingness to defer gratification, to name just three – is it acceptable to use this information to discriminate between employees? The company decided that it wasn't, in part because it would have been unfair to less privileged applicants, who may have had to use a library computer to apply.

Another problem with data models is that they may suffer from a psychophysics problem. They match reality with behaviour as though the one maps perfectly onto the other. But this is wrong. For example, the data might say that people won't pay £49 for a jar of coffee and that's true, mostly. However people will pay 29p for a single Nespresso capsule which amounts to a similar cost – without understanding human perception it is unable to distinguish between the two. Will people pay £100 for a pair of shoes? In Walmart, no chance, but in the designer store Neiman Marcus, easily. Will people pay £500 for a mobile phone handset? Nokia's data said no, but Apple discovered they would. Big data makes the assumption that reality maps neatly on to behaviour, but it doesn't. Context changes everything.

Perception may map neatly on to behaviour, but reality does not map neatly onto perception.

We should also remember that all big data comes from the same place: the past. Yet a single change in context can change human behaviour significantly. For instance, all the behavioural data in 1993 would have predicted a great future for the fax machine.

6.5 NOTHING NEW UNDER THE SUN

It is arguable that even the ancient Greeks grasped the principles of psychophysics. Study the Parthenon, and you'll notice that there is barely a straight line in it; the floor curves upwards in the middle, the sides bow out and the columns swell in the middle.* This is because it is not designed to be perfect – it's designed to *look* perfect to a human standing a hundred yards or so downhill. And long before the Parthenon, nature learned the same trick.

Nature spends a great deal of resources on what might be called 'perception hacking' or, in business terminology, marketing. Berries and fruits that want to be eaten develop a distinctive colouration and an attractive taste when they ripen. By contrast, caterpillars that don't want to be eaten have evolved to taste disgusting to their predators. And some butterflies produce what look like eyes on their wings because many animals react more cautiously in their presence. Such are examples of how nature is able to hack perception rather than changing reality.

* A trick borrowed by the designer of the Rolls-Royce radiator grille. 'Entasis' is the technical term.

6.6 WHEN IT PAYS TO BE OBJECTIVE – AND WHEN IT DOESN'T

If you are a scientist, your job is to reach beyond the quirks of human perception and create universally applicable laws that describe objective reality. Science has developed sensors and units of measurement, which measure distance, time, temperature, colour, gravity and so on. In the physical sciences we quite rightly prefer these to warped perceptual mechanisms: it does not matter whether a bridge *looks* strong – we need to know that it really is strong.

A problem arises when human sciences – politics, economics or medicine, say – believe this universalism to be the hallmark of a science and pursue the same approach; in the human sciences, just as in TV design, what people *perceive* is sometimes more important than what is objectively true. In medicine, the obsession with objectivity leads to neglect of the placebo effect when it is seen as a 'mere' perceptual hack. But if a treatment like homeopathy, say, leads people to believe they will get better, and this happy delusion helps them feel less ill, then what's not to like? Surely we should be researching this rather than decrying it?

But what about snake-oil salesmen,* the fakers, fraudsters and conmen? Alchemy, precisely because it is not an exact science, has

* First of all, it is not quite fair to excoriate all snake-oil salesmen – in the days before antibiotics, snake oil

always been rife with charlatanry, and we should be on our guard for this. Many of the remedies proposed by people in advertising and design are wrong, and many of the findings of behavioural scientists have already been or will be proved wrong. Some parts of this book are also undoubtedly wrong – I am conscious that I have written this book from an incredibly optimistic perspective, but my argument is not that alchemy is always reliable, ethical or beneficial. Far from it – it is simply that we should not recoil from testing alchemical solutions because they do not fit with our reductionist ideas about how the world works. The purpose of this book is to persuade the reader that alchemy exists whether we like it or not, and that it is possible to use it for good; besides, if people are more aware of its existence, they will be better at spotting its misuses.†

In physics and engineering, objective models usually make problems easier to solve, while in economics and politics objectivity might make things harder. Some pressing economic and political issues could be solved easily and cheaply if we abandoned dogmatic universal models; just as TV designers don't wrestle with the problem of producing the entire spectrum of visible light, policymakers, designers and businessmen would be wise to spend less

was quite possibly the best solution available. Genuine snake oil, from the fat of the Chinese water snake, contains 20 per cent eicosapentaenoic acid, which has potent analgesic and anti-inflammatory properties and was used successfully in Chinese medicine for centuries. However, the substance more commonly known as snake oil was a series of concoctions, often high in alcohol or opiates, that claimed to contain snake oil. Typically they might contain a mixture of herbs, possibly to give a plausibly strange taste.

† There are quite a few misuses of alchemy, which I believe should be outlawed. The insanely low 'minimum payment' for credit cards is a black-magic trick that encourages debt, for instance.

time trying to improve objective reality and more time studying human perception and moral instinct.

Every day, companies or governments wrongly make highly simplistic assumptions about what people care about. Two major US retailers, JCPenney and Macy's, both fell foul of this misunderstanding when they tried to reduce their reliance on couponing and sales, and instead simply reduced their permanent prices. In both cases, the strategy was a commercial disaster. People didn't want low prices – they wanted concrete savings. One possible explanation for this is that we are psychologically rivalrous, and like to feel we are getting a better deal than other people. If everyone can pay a low price, the thrill of having won out over other people disappears; a quantifiable saving makes one feel smart, while paying the same low prices as everyone else just makes us feel like cheapskates. Another possible explanation is that a low price, unlike a discount, does not allow people any scope to write a more cheerful narrative about a purchase after the event – 'I saved £33', rather than 'I spent £45'.

It is worth remembering that costly signalling may also play a role in this: certain things need to be expensive for symbolic reasons. A £200 dress reduced to £75 is fine, but women may not feel happy wearing a £75 dress to a wedding. TK Maxx,[‡] a psychologically ingenious retailer, is a brilliant place to buy a present for your wife, provided that under no circumstances you reveal to her that you bought it from TK Maxx.[§]

Economic logic is an attempt to create a psychology-free model of human behaviour based on presumptions of rationality, but it can be a very costly mistake. Not only can a rational approach to pricing be very destructive of perceived savings, but it also assumes that everyone reacts to savings the same way. They don't, and

‡ It is, of course, known as TJ Maxx in the US.

§ A £500 dress feels like a £500 dress when you wear it, even if you bought it for £200 (I am relying on my wife not making it this far into the book).

context and framing matter. One of my favourite ever experiments on the perception of price and value came from the father of 'Nudge Theory', Richard Thaler. He asked a group of sophisticated wine enthusiasts to imagine that they had bought a bottle of vintage wine (that was now worth $75) some years before for $20. Then he asked them to choose an answer that best reflected the cost to them of drinking the bottle:

a. $0. I already paid for it: 30 per cent
b. $20. What I paid for it: 18 per cent
c. $20 plus interest: 7 per cent
d. $75 – what I could get if I sold the bottle: 20 per cent
e. -$55. I get to drink a bottle that is worth $75 that I only paid $20 for, so I save money by drinking this bottle: 25 per cent

These results reveal that *some* people do think like economists – but they seem to be a 20 per cent minority.

Notice, too, that they are the people who will enjoy the wine least. (There's a reason it's called 'the dismal science'.)

6.7 HOW WORDS CHANGE THE TASTE OF BISCUITS

Remember that words do not only affect the price of a dish – they can also change its taste. Five years ago, we received a worried phone call from Belgian colleague. One of their largest biscuit manufacturers had replaced their most popular brand with a new, lower-fat variant, but as soon as they released it onto the market, sales plummeted. They were bamboozled; they had performed extensive research and testing and many people could notice no difference in the taste of the new biscuit, and yet no one was buying the new version.

This was one of those problems which I was able to solve without even leaving my chair. 'I see,' I said over the speakerphone, 'And did you put "Now with lower fat" on the packaging of the biscuits?' 'Of course we did!' they replied. 'We'd spent months reducing the fat content of the biscuits – what's the point of doing that if you don't tell people about it?' 'There's your problem,' I said. 'It doesn't matter what something tastes like in blind tastings, if you put "low in fat" or any other health indicators on the packaging, you'll make the contents taste worse.' In the testing that they had conducted the biscuits had been unpackaged, and they had forgetten that packaging will also affect the taste.

6.8 THE MAP IS NOT THE TERRITORY, BUT THE PACKAGING IS THE PRODUCT

The Polish-American academic Alfred Korzybski (1879–1950) is perhaps most famous for his dictum that 'The map is not the territory.' He created a field called general semantics, and argued that because human knowledge of the world is limited by human biology, the nervous system and the languages humans have developed, no one can perceive reality, given that everything we know arrived filtered by the brain's own interpretation of it. Top man!

One day, Korzybski offered to share a packet of biscuits, which were wrapped in plain paper, with the front row of his lecture audience. 'Nice biscuit, don't you think?' said Korzybski, while the students tucked in happily. Then he tore the white paper and revealed the original wrapper – on it was a picture of a dog's head and the words 'Dog Cookies'. Two students began to retch, while the rest put their hands in front of their mouths or in some cases ran out of the lecture hall and to the toilet. 'You see,' Korzybski said, 'I have just demonstrated that people don't just eat food, but also words, and that the taste of the former is often outdone by the taste of the latter.'

This effect is not confined to edible products; in cleaning products, adding the words 'now kinder to the environment' to packaging may lead people to instinctively believe the contents are less effective. There is an ethical and practical dilemma here:

if you wish to produce a greener laundry detergent, should you mention its kindness to the environment on the packaging, given that those claims might unconsciously cause people either to not to buy it or to use more of it than is necessary? In some cases it might be better to do good by stealth, particularly if the number of buyers who care about the environment may be outnumbered by those who don't.

Announcing even the tiniest tweaks to popular products has been a disaster for Vegemite, Milo and the Cadbury Creme Egg – even if they couldn't otherwise, people notice a change in taste simply because a change in formulation has been announced. When Kraft wanted to introduce a healthier formulation for their Mac & Cheese, they were terrified of a similar reaction, particularly as the malignant combination of social media and newspapers keen for a story can turn a small number of hostile tweets into national news. So they removed the artificial yellow dye and added paprika, turmeric and other natural replacements – and then kept silent about it. Practically no one noticed a thing – until they announced the change retrospectively, under the headline 'It's changed. But it hasn't.' This way, they were able to gain the potential custom of people who had previously avoided the product because of its artificiality, without creating an imagined taste change among its regular customers, who suddenly discovered they had been eating the healthier variant all along.*

---

* Congratulations to the agency alchemists at Crispin Porter + Bogusky for providing this magic. It can't have been easy to sell to the client – after all, why do something beneficial and keep it quiet?

6.9 THE FOCUSING ILLUSION

Attention affects our thoughts and actions far more than we realise. Daniel Kahneman, along with Amos Tversky, is one of the fathers of behavioural economics; 'the focusing illusion', as he calls it, causes us to vastly overestimate the significance of anything to which our attention is drawn. As he explains:

'Nothing is as important as we think it is while we are thinking about it. Marketers exploit the focusing illusion. When people are induced to believe that they "must have" a good, they greatly exaggerate the difference that this good might make to the quality of their life. The focusing illusion is greater for some goods than for others, depending on the extent to which the goods attract continued attention over time. The focusing illusion is likely to be more significant for leather car seats than for books on tape.'

In marketing, comparison tables can be used to play a trick on the consumer: if the person compiling the table in the above advert for car breakdown services had wanted to be objective, there are perhaps 50 additional benefits he could have added which are offered by all the companies. However, he instead chooses to focus the reader's mind on the small proportion of overall benefits that are uniquely offered by the brand being promoted.

The old advertising belief in having a Unique Selling Proposition (a 'USP') also exploits the focusing illusion: products are easier to

	ETA	GREEN FLAG	GREEN INSURANCE COMPANY	GEM	AA	RAC
Free parts and labour cover?	✓	✗	✗	✗	✗	✗
Free second recovery?	✓	✓	✗	✗	✗	✗
Misfuelling covered?	✓	✓	✓	✗	✗	✗
Quick recovery pledge?	✓	✓	✓	✗	✗	✗
Credit for cancelled vehicles?	✓	✓	✓	✗	✗	✗
Carbon emissions offset?	✓	✓	✓	✗	✗	✗
Maximum callouts per year	Unlimited	Unlimited	6	Unlimited	7	6
Maximum passengers recovered	Legal carrying capacity	Legal carrying capacity	9	8	7	6

Emphasise the differences, not the similarities.

sell if they offer one quality that the others do not. Even if this feature is slightly gratuitous, by highlighting a unique attribute, you amplify the sense of loss a buyer might feel if they buy a competing product.

Camping equipment is one of the most dangerous things to buy while in the grip of the focusing illusion. While in the shop, you imagine yourself using the products in perfect climatic conditions, but these in fact rarely occur.* Secondly, the facets of a product that seem most appealing at the moment of purchase may, in fact, be disadvantageous when we come to use them. For example, all sleeping bags when sold are packed tightly into unfeasibly small bags. However, the long-term effect of these is that, though the product may look attractive when new and professionally packaged, they are all but impossible to repack after use.

* In my homeland, anyway.

We can guard against this illusion by directing our attention to metrics that may seem less salient than they deserve to be. For example, we can imagine ourselves trying to repack a tent in the rain on a windy day. Or, while looking to buy a Porsche, we perhaps should imagine ourselves sitting in it while stuck in London traffic – something that will happen many times – rather than on a drive through the Cotswolds on a summer's evening – something that will happen only once or twice.

It is interesting that Kahneman uses leather seats in cars as an example of the focusing illusion. I had always wondered whether they were chosen purely because of status signalling: status probably ranks higher on your list of priorities when you buy a car than it does later on in the car's life, when reliability, running costs and comfort are more significant. Actually, whether leather seats are better than cloth seats *also* depends on the focusing illusion. There are many dimensions by which you can compare what seats are covered in, including not only status and price, but slippiness, smell, ease of cleaning, durability, ethics and even whether they can be painful to sit on in hot weather.† Depending on which of these qualities you choose to use as a discriminator, leather is either vastly superior to cloth or a senseless extravagance.‡

It is fair for Kahneman to say that the focusing illusion plays a huge part in marketing, but I would argue that it is not actually an

†† I am told that anyone who has had a child be sick in a car with cloth seats immediately becomes a fan of leather ones.

‡‡ It is worth remembering that Professor Kahneman is an academic, and hence a member of a strange caste where having a crappy car seems to be a badge of honour ('countersignalling' is the technical term for this). If you want to destroy your academic career in the social sciences on day one, turn up in the university car park in a brand new Ford Mustang (a vintage one might be acceptable at a pinch, but only once you have reached the rank of tenured professor).

illusion at all but is rather an evolutionary necessity. Furthermore, rather than marketers 'exploiting' the focusing illusion, it is the illusion which makes marketing necessary. Nevertheless, one way you can improve your happiness is by learning that such an illusion exists, and by controlling what you pay attention to. I have a soft spot for the religious practice of saying grace before a meal, since paying attention to good things that one might easily take for granted seems a good approach to life – a pause to focus attention on a meal should add to its enjoyment.§

§ Perhaps the modern equivalent of saying grace is photographing your food.

6.10 BIAS, ILLUSION AND SURVIVAL

The focusing illusion is indeed an illusion, but so is almost *all* our perception, because an objective animal would not survive for long. As neuroscientist Michael Graziano explains, 'If the wind rustles the grass and you misinterpret it as a lion, no harm done. But if you fail to detect an actual lion, you're taken out of the gene pool.'* It is thus in our evolutionary best interests to be slightly paranoid, but it is also essential that our levels of attention vary according to our emotional state. When walking on our own down an unlit street, the sound of footsteps will occupy more of our attention than it would on a crowded street in daylight.

It is wrong to consider such illusions as things that should be corrected or avoided – it is worth understanding them and the role they might play in distorting our behaviour, but the idea that it is better not to experience them is highly dangerous. For example, if your smoke alarm possessed consciousness, you might confidently say that it suffered from paranoid delusions – it might start beeping furiously when you were doing nothing more dangerous than making toast. There's a good reason for this – it cannot easily distinguish between the early stages of a house fire and a slice of

---

* 'A New Theory explains How Consciousness Evolved', *Atlantic* (6 June 2016).

burnt toast. However, the consequences of being wrong in this situation are quite different. If the smoke alarm goes off when you burn your breakfast, it's a false positive and annoying, but a false negative can be fatal; the last kind of smoke detector you would want is one that only activates when the flames are licking at its edges.

We have to be careful before we start to casually label biases and illusions as inherent mental failings, rather than the product of

Pareidolia in action – some people see a face, while some even see George Washington.

evolutionary selection. It pays to consider pareidolia, an optical illusion from which many of us 'suffer' when we see faces and human or animal shapes in inanimate objects.

You don't have to study much evolutionary biology to work out why we are highly attuned to detecting faces or animals in the environment. Many threats in our evolutionary history would have been posed by other animals, and being able to recognise them and read their mood would often have made the difference between life and death. Just as with smoke detectors, this raises a question of calibration – yes, the price you pay for being good at spotting

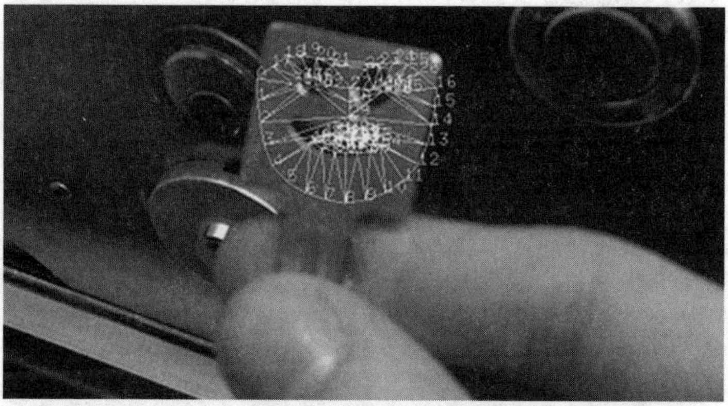

In the photograph of the key, an algorithm picked out a human face from what is merely two dots and a line cut into it. Software suffers from pareidolia too.

human or animal faces is that you tend to see them when they aren't really there, but it is a price worth paying. It may mean that you might attribute emotions to your washing machine or believe there is a human face in a rock formation, but this comes at little cost to your evolutionary fitness, compared to the advantages that a heightened talent for facial recognition brings.

A brain which saw faces in *every* rock or tree, however, would not be very useful. We can assume that, if possible, evolution will seek to improve its calibration and reduce the number of unwanted false positives. Both smoke detectors and car alarms are significantly less paranoid today than they were in the 1980s but they are both calibrated to err on the side of caution, which is the way it should be. There is always a trade-off, and illusions are the price we pay.

In the same way, facial recognition software must make the same trade-off in order to work. If it *never* recognises a face mistakenly, it is too insensitive – such high standards would mean that it would fail to recognise faces at a slight angle, or if one eye were closed, which would render it useless. As a result, facial recognition algorithms experience exactly the same pareidolic illusions as humans and some form of calibration is always necessary when dealing with imperfect or ambiguous information. What this all means is

A 'confused washing machine' – another example of pareidolia.

that no living creature can evolve and survive in the real world by processing information in an objective, measured and proportionate manner. Some degree of bias and illusion is unavoidable.

6.11 HOW TO GET A NEW CAR FOR £50

Do you own a car? If you do, is it a reasonably good one? If it is, I have good news for you: the next paragraph will earn you the price of this book back, many times over.

The next time you are thinking about replacing your car, don't. Instead, wait at least a year, or maybe two or three. In the meantime, rather than selling it, take it to a good valet service from time to time and have it thoroughly cleaned, inside and out. This will cost you about £50–£100 each time, but you will have a much better car. Not just a cleaner car, but a *better* car – as well as looking nicer, it will drive more smoothly, accelerate more quickly and take corners more precisely. Shiny cars are also simply much more enjoyable to drive. Why? Because of psychophysics.

6.12 PSYCHOPHYSICS TO SAVE THE WORLD

How can you stop environmentally friendly cleaning products from being perceived as less effective? Fortunately, there are tricks you can play to fool your unconscious into thinking that environmental gains don't necessarily come at the cost of effectiveness. Again, such tricks fall under the category of 'benign bullshit'.

One way in which businesses can reduce their environmental footprint is to sell a product in concentrated form, which reduces packaging and distribution costs, and can also reduce the volume of chemicals used. But there are several problems:

1. Some people will continue to use the same volume as before, despite the increase in concentration, which leads to overdosing. A smaller cap might reduce this problem, but some people cannot help assuming that less volume means less effect, and compensate with an extra capful.
2. People may not buy the product at all, because although more concentrated it looks to be worse value for money on the shelf.
3. People may believe that the product is inferior simply because there is less of it, and lose faith in its value.
4. A smaller product takes up less space on the shelf, which may reduce its visibility and allow more room for the products of competitors.

There are a few ways to counteract this:

1. Radical honesty,* such as announcing that the product is, say, 4 per cent less powerful than previously, but 97 per cent better for the environment. Or alternatively, be explicit about a product's weakness.†
2. Deploy the 'Goldilocks effect' – the natural human bias that means that, when presented with three options, we are most likely to choose the one in the middle. Washing detergent manufacturers use language that normalises the lower and middle usage of the product, while implicitly stigmatising overdosing. For example, 'Half a capful for light–normal wash'; 'One capful for a full or heavy wash'; 'Two capfuls for extreme soiling.' This creates the impression that one would only use more than one capful if they had committed some brutal crime: as a result, even overdosers will likely use only one cap.‡
3. Change the format: it is hard to believe that a lower amount of powder or liquid will do the same job as before, but if the formulation is changed to a gel or tablets we are

* No company as far as I know has adopted this approach, I suspect because it would be difficult to sell within an organisation.

† An admission of a downside makes a claim more plausible. Great advertising taglines often harness this effect – examples include 'Reassuringly expensive' and 'We're number two so we try harder.'

‡ The 'Goldilocks effect' can equally be used by the manufacturers of washing machines who may wish for environmental reasons wish to encourage the use of lower temperature cycles. If you offer a couple of very low temperatures, so that 30 and 40 degrees lie in the middle of the knob, with 60 and 90 degrees to the extreme right, then people will instinctively gravitate towards the lower temperature washes.

more likely to believe it. And if you produce the product in tablet form, consider packaging them in a thin, wide and high packet, so their visibility on the shelf is not reduced.

4. Add intricacy: simply adding coloured flecks to a plain white powder will make people believe it is more effective, even if they do not know what role these flecks perform. Similarly, tablets that consist of a mixture of liquids, gels and powders help people believe that less is doing more. Remember stripy toothpaste.
5. Add effort. If a concentrated product requires you to mix it with water first, or to mix together two separate ingredients before using it, our belief in its potency is restored by this small amount of extra bother.[§]

All these solutions seem like bullshit from a logical point of view, and they do all involve an element of smoke and mirrors. If we were capable of objectively viewing the world we would regard this as deception, but alas we can't. Also, it's not as if, without these smoke and mirrors, we'd suddenly see the world with perfect accuracy – we'd just see different smoke and different mirrors.

So, do you want the smoke and mirrors that help the environment, or the smoke and mirrors that don't?

§ I was intrigued to discover that a client's most successful concentrated product (a wood polish) was the one product where its use required an extra stage of effort, in this case diluting it in a small container.

6.13 THE IKEA EFFECT: WHY IT DOESN'T PAY TO MAKE THINGS TOO EASY

In the 1950s, the General Mills food company launched a line of cake mixes under the Betty Crocker brand that included all the dry ingredients, including milk and eggs. All you needed to do was add water, mix and stick the pan in the oven – what could go wrong? However, despite the many benefits of this miracle product, it did not sell well, and even the Betty Crocker name could not convince anyone to buy it. General Mills brought in a team of psychologists to find out why consumers were avoiding it. One of their explanations was guilt: the product was so damned *easy* to make compared to traditional baking that people felt they were cheating. The fact that the cake tasted excellent and received plaudits didn't help – this simply meant that the 'cook' felt awkward about getting more credit than they had earned.

In response to these results, General Mills added a little psychological alchemy – or 'benign bullshit'. They revised the instructions on the packaging to make baking less convenient: as well as water, the housewife was charged with adding 'a real egg' to the ingredients. When they relaunched the range with the slogan 'Just Add an Egg', sales shot up. The psychologists believed that doing a little more work made women* feel less guilty, while still saving time,

* This was the 1950s, remember.

but making just enough effort to give the sense of having contributed to the cake's creation.

There is a name for the addition of consumer effort to increase someone's estimation of value. It should perhaps be called the Betty Crocker effect, since they spotted it first, but it's instead known as the IKEA effect, because the furniture chain's eccentric billionaire founder Ingvar Kamprad was convinced that the effort invested in buying and assembling his company's furniture added to its perceived value. When working with IKEA I was once advised: 'Do not, under any circumstances, suggest ways of making the IKEA experience more convenient. If you do, we shall fire you on the spot.'

We employed this effect a few years later, when asked to help promote a fabric detergent which had been designed for the developing world – it required clothes to be rinsed once, rather than three times during washing, in order to save water. Our idea was to create a more complex bucket to replace the three buckets that had been required previously, which would add a degree of gratuitous complication to the single rinse. This improvement to the detergent's efficacy was only slight: the real point of the extra effort involved was to prevent the new process from seeming too good to be true.

A final note. When working with pharmaceutical companies, I discovered that every developer tried to make their drug as easy to ingest as possible – however, the behavioural economist Dan Ariely and I disagree with this apparently logical assumption. We both feel that the placebo effect might be strengthened if the drug requires some preparation, whether prior dilution or mixing. In additon, by creating a routine around the preparation of a drug before you take it you also create a ritual, which makes it much harder to forget. It's easy to forget whether you have swallowed two miniscule pills, but much harder to forget whether you have mixed liquid A with liquid B before adding powder C.

6.14 GETTING PEOPLE TO DO THE RIGHT THING SOMETIMES MEANS GIVING THEM THE WRONG REASON

As I mentioned earlier, the human brain to some extent automatically assumes that there are trade-offs in any decision. If a car is more economical, its performance is assumed to be more sluggish; if a washing powder is kinder to the environment, it is assumed to be less effective. This is why promoting a product as being 'kind to the planet' comes at a risk – might it be easier to save the planet if we talked less about doing so? The error of the environmental movement seems to me to be assuming that it is not only necessary for people to do the right thing, but that they must do the right thing *for the right reasons*. My own view is more cynical, and also pragmatic: if people adopt behaviours that benefit the environment, we shouldn't really care what their motives are.

Demanding people do the right thing *and for the right reason* is setting the bar rather too high. When Ogilvy was asked to increase the level of waste recycling in British homes we made the suggestion of shelving all discussion of what a household thinks about the growth of landfill or the loss of polar bears; instead we suggested that the principal behavioural driver of recycling is to do with circumstances rather than attitude. Put bluntly, if you have two bins in your kitchen, you'll separate your recyclable rubbish and recycle quite a lot, but if you have only one bin you probably won't. Under the slogan 'One bin is rubbish' we focused our campaign entirely around encouraging people to have more than one bin in their household – avoiding the issue of how to convert people to be card-carrying members of the green movement.*

We were not being defeatist in this campaign, or giving up on the attempt to make people more environmentally aware – we were just solving the problem backwards. Conventional wisdom about human decision-making has always held that our attitudes drive our behaviour, but evidence strongly suggests that the process mostly works in reverse: the behaviours we adopt shape our attitudes. Perhaps someone who separates their rubbish into waste and recyclables will become more environmentally conscious as a result of having adopted the behaviour, just as Tesla drivers will wax enthusiastically about the environmental purity of their vehicles, regardless of their initial reasons for buying the car.†

Behaviour comes first; attitude changes to keep up.

We have adopted a similarly pragmatic approach in proposals to reduce the amount of uneaten supermarket food thrown out by consumers once it passes a best-before date. Again, we didn't concentrate on the reasons people shouldn't waste food, but instead on ways to make unwasteful behaviour easier to adopt. Our suggestions have included such childishly simple solutions as including the day of the week on 'Use By' and 'Best Before' dates

---

* We are also developing a free plastic clip that allows people to attach a second bin-bag to the outside of their existing bin.

† Let me make a prediction right here: very few people who buy a Tesla will ever go back to owning a conventional car, because the act of buying it will affect their preferences indelibly. But this lasting change in behaviour will not necessarily have been instigated by care for the environment, any more than people who installed indoor lavatories and baths did so to minimise the risk of a cholera outbreak.

on packaging. 'Use by Friday, 12/11/17' is a much more useful reminder than a numerical date.‡

As we have seen in this section, it is *only* the behaviour that matters, not the reasons for adopting it. Give people a reason and they may not supply the behaviour; but give people a behaviour and they'll have no problem supplying the reasons themselves.

‡ 'Friday' is, to use the terminology of Daniel Kahneman, much more 'System 1 friendly' than '12/11/17', requiring less cognitive effort to deliver its meaning.

PART 7: HOW TO BE AN ALCHEMIST

7.1 THE BAD NEWS AND THE GOOD NEWS

After landing at Gatwick Airport, the plane taxied for five minutes or so before coming to a halt, the terminal still somewhere in the distance. I heard the engines wind down and a horrible thought occurred to me: we might be about to be loaded onto a bus. I had always felt mildly resentful about being bussed to the terminal, suspecting that it was a tactic used by airlines to save landing fees by parking far away from the terminal building and avoid paying for an airbridge.

Then the pilot made an announcement that was so psychologically astute that I felt like offering him a job at Ogilvy. 'I've got some bad news and some good news,' he said. 'The bad news is that another aircraft is blocking our arrival gate, so it'll have to be a bus; the good news is that the bus will drive you all the way to passport control, so you won't have far to walk with your bags.'

After decades of flying, I suddenly realised that what he had said was not just true on that occasion – it was *always* true! The bus drops you off exactly where you need to be, meaning that you don't have to lug your carry-on bags through miles of corridors before you can get to the exit – this was a revelation. We soon arrived at passport control, and were all rather grateful for the bus. Nothing had changed objectively, but we now saw the bus not as a curse but as a bit of a bonus. The pilot's alchemical approach had redirected my attention to a different judgement.*

* You can try to perform the same alchemy the next time you are on a flight. If a bus appears to take you to the terminal, say out loud to your companion that you are glad, since it will drive you all the way to passport control. You will have increased the happiness levels of everyone within earshot.

7.2 ALCHEMY LESSON ONE: GIVEN ENOUGH MATERIAL TO WORK ON, PEOPLE OFTEN TRY TO BE OPTIMISTIC

One characteristic of humans is that we naturally direct our attention to the upside of any situation if an alternative narrative is available, minimising the downside. By giving people good news and bad news at the same time, you can make them much happier than they would be if left with only one interpretation: that pilot was perhaps cleverer than he knew.

One of most amusing and telling stories in the recent history of behavioural economics appears in Richard Thaler's memoir, *Misbehaving* (2015), when he describes what happened when the University of Chicago Economics Faculty was required to move to a new building. These people are, in theory, the most rational in the world, who should have had at their disposal every possible strategy for collective decision-making in allocating the offices, which varied slightly in size and in status (a corner office being more desirable than an office with one window). Some of them suggested an auction, but this idea was quickly rejected – it was regarded as unacceptable for elderly Nobel Laureates to have smaller offices than their younger colleagues just because the latter had lucrative consulting practices and could afford to pay for the best offices.

There was a great deal of feuding and obsessing over tiny differences in size.*

I suggested to Professor Thaler that there might have been a simpler way to solve the problem by using a little psychological alchemy: why not rank both the offices and the faculty parking spaces between 1 and 100 in order of desirability before allocating them by lottery, with the people who received the best offices receiving the worst parking spots, and vice versa? Under these conditions, people allocate greater importance to the part of the lottery in which they have come out best, while those in the middle reframe the result as being a happy compromise.

I was familiar with this system because it was how rooms had been allocated at my university college, a practice that I believe has gone on for centuries. In their first year, everyone is allocated a fairly standard room in college – none of the markedly good ones are given to first-year students. In their second year, a ballot is run, with the person at the top receiving the first pick, the second the second, and so on; and in the third year, the positions on the ballot are reversed. I never met anybody who was unhappy with the result.

This seems to offer an extraordinarily valuable psychological insight into the best way to divide unequal resources between a random collection of people – when presented with either good plus bad, bad plus good or average plus average, everybody seems equally content. In fact, we seem well-disposed to explicit trade-offs. A sentence which contains bad and good news, along the lines of, 'Yes we admit downside x, but also think of upside y,' seems particularly persuasive. Robert Cialdini has observed that, as you are closing a sale, the admission of a downside oddly adds

* As Thaler remarks, the argument was rather pointless, since all the offices were of a perfectly adequate size; moreover, the offices on the less fashionable side of the building had the compensation of a view of Robie House, one of Frank Lloyd Wright's Prairie-style masterpieces.

persuasive power: 'Yes, it is expensive, but you'll soon find it's worth it,' seems to be a strangely persuasive construction – explicitly mentioning a product's weakness enables people to downplay its importance and accept the trade-off, rather than endlessly worrying about the potential downside. If you are introducing a new product, it might pay to bear this in mind.

When you think about it, it is rather strange how explicit low-cost airlines are about what their ticket prices *don't* include: a pre-allocated seat, a meal, free drinks, free checked luggage – such deficiencies help to explain and destigmatise the low prices. 'Oh, I see,' you can say, when you see a flight to Budapest advertised for £37, 'the reason that low price is possible is because I won't be paying for a lot of expensive fripperies that I probably don't want anyway.' It's an explicit, well-defined trade-off, and one that we feel happy to accept.

Imagine if cheap airlines instead claimed: 'We're just as good as British Airways, but at a third of the price.' Either nobody would believe them, or else such a claim would raise instant doubts: 'Maybe the only reason they're cheaper is because they don't bother servicing the engines or training the pilots, or because the planes are scarcely airworthy.'

So, marketing can not only justify a high price but it can also detoxify a low one. Make something too cheap without sufficient explanation and it simply might not be believable – after all, things which seem too good to be true usually are.

7.3 SOUR GRAPES, SWEET LEMONS AND MINIMISING REGRET

My airport bus transfer experience illustrates a simple truth about human psychology that was spotted over two thousand years ago by a wise storyteller called Aesop. It appears in several of his fables including, most famously, the story of the fox and the grapes. A fox gazes longingly at a beautiful bunch of grapes hanging from the high branch of a tree, his mouth watering. He jumps to reach them, but misses by a long way. He tries again and again, but each time it is in vain. Sitting down, he looks at the grapes in disgust. 'What a fool I am,' he says, 'wearing myself out to get a bunch of sour grapes* that are not worth the effort.'

The moral of the fable is that many pretend to despise and belittle that which is beyond their reach. This seems fair enough, though it is worth asking how our lives would feel if we did not play this mental trick on ourselves – we might go about in a constant state of resentment because we were not the billionaire recipient of a Nobel Prize.

The opposite phenomenon to sour grapes is often called 'sweet lemons', where we 'decide' to put a positive spin on a negative expe-

* Which is where the phrase 'sour grapes' comes from – it must surely be one of the oldest metaphors in continuous use.

rience. Both these mental tricks are types of 'regret minimisation' – given the chance, our brain will do its best to lessen any feelings of regret, though it does need a plausible alternative narrative to do this. Thinking back to my experience at the airport, the reason I had previously hated being bussed to the terminal was not because it was intrinsically bad, but because there didn't seem to be anything that would help me frame it in a more positive light. Once I knew of an upside, I was able to choose to see the bus as a conveyance and not an annoyance. As Shakespeare wrote, 'there is nothing either good or bad, but thinking makes it so'.

A few hours before I sat down to write this chapter, I received a parking ticket. It was only for £25 and I was completely to blame, but it nevertheless annoyed me to an extraordinary degree – and it is still annoying me now. Perhaps a parking ticket is made even more annoying because we can see no way of reframing it in a positive light.

Could the local authority that issued me with the ticket give me a chance to play the same mental trick on myself as the easyJet pilot – a reason, however tenuous, to feel slightly upbeat about the fine? For instance, how different would I feel if I was told that the money from my fine would be invested into improving local roads or donated to a homeless shelter? The fine would have the same deterrent effect, but my level of anger and resentment would be significantly reduced. How would that be a bad thing?

7.4 ALCHEMY LESSON TWO: WHAT WORKS AT A SMALL SCALE WORKS AT A LARGE SCALE

Why don't we take the insight gained from our experiences of easyJet and parking tickets and apply it to something larger? Public services are sometimes disliked by the people who use them not because they are worse than private sector services, but because the link between what you pay and what you gain is so opaque that it prevents people from creating a positive narrative about the taxes they pay.*

I once examined a breakdown of what my local taxes are spent on: it seemed that I paid £25 per year for weekly refuse collection – reframing this as 50p per week, I was struck by what good value it was. For less than the cost of a stamp, someone would come to my house and remove and dispose of several bags of refuse; suddenly the council seemed more impressive than they had before.

One problem with governments is that they generally hate hypothecation, the system whereby taxes are ring-fenced and spent on a pre-defined area of activity. Instead, taxation tends to

* Countries such as Denmark and Sweden may be an exception to this: in both countries the tax rates are egregiously high, but the public expenditure is subject to a high level of democratic and local scrutiny.

go into one pot, before being spent wherever it is needed. As a result, we end up resenting taxation more than expenditure that goes towards something we can see, feel or even imagine.

By contrast, consider for a second the success of private organisations in securing philanthropic donations when they can offer the donor something in return – even something as trivial as naming rights to a building. The taxation we give to the government doesn't provide us with the opportunity to create a narrative that might make us happy about what we pay. Taxes, like parking fines, are seen as wholly bad, but a little alchemy could solve this problem quite easily. In ancient Rome, wealth taxes were levied to fund military campaigns or public works, and since the names of the people who paid them were displayed on a monument, with the money dedicated to a specific end, rich people were happy to pay, with those initially deemed too poor to be liable volunteering themselves, saying 'Actually I'm much richer than you think.'

People who are happy to spend £300 on a pair of designer sunglasses[†] feel resentful when they are required to pay that amount on funding healthcare, policing, the fire brigade or defence, yet many of us would voluntarily pay more tax if it were possible to specify which service would receive the money.[‡] If you allowed people to tick a box on their income tax form whereby they paid 1 per cent extra to improve healthcare, many people would happily do so. And if you then gave them a car sticker to display the fact that they had paid more, as the Romans effectively did, even more people would join in. How would that be a bad thing? And yet, for some reason, government institutions and businesses seem to shy away from such solutions. Perhaps they think of them as cheating; maybe they *are* cheating, but the fact remains that, if emotional

† An item that probably costs little more than £15 to manufacture.

‡ I am not alone in thinking this – Shlomo Benartzi from the University of California, Los Angeles, recently proposed something similar to the UK Government.

responses have such an influence on our brain, we have no choice but to at least try to present things in a way that it finds least emotionally painful.

Remember, no one will buy a fish, however tasty, if it's called the Patagonian toothfish.

And similarly, perhaps no 26-year-old will ever buy a financial product, however advantageous it may be to them, if it is called a pension. At present the UK Government spends over £25 billion each year on tax rebates for pension contributions, a staggeringly generous incentive to save for retirement by any standards, and yet it is staggeringly ineffective. I was recently part of a group that met to discuss what the government could do to make paying into a pension more appealing, particularly to younger people, without requiring such a high level of financial subsidy. We were all impressed by the work that Richard Thaler and Shlomo Benartzi had already performed in this field: together, they conceived a new mechanism for pension-saving that acknowledges one of the central principles of behavioural psychology – loss-aversion, the mental mechanism that causes us to experience more pain from losing £100 than pleasure from winning £100.[§]

A typical pension works like this: if you buy a pension plan for £250 a month, every month thereafter you are £250 poorer, until your retirement, when you can redeem the annual salary which

§ See 'Save More Tomorrow', *Journal of Political Economy* (February 2004).This runs counter to the economic idea of utility, and is often considered to be an example of human irrationality (a 'bias') by economists. However, it may be that our brains have evolved to be right and that economists' assumptions of rationality are wrong. In a path-dependent life, a loss (or, worse still, a series of losses) would bring a higher risk of damaging consequences than an equivalent gain would have benefits. Two or three consecutive losses in quick succession could easily make the difference between survival and extinction.

that pension provides. By contrast, Thaler and Benartzi's 'Save More Tomorrow' pension worked differently: you signed up for a pension at a certain rate (let's say 20 per cent) but instead of starting immediately, your contributions would only represent a proportion of any future wage *rises*. So if you were given a £500-a-month pay rise, 20 per cent of it (if that is what you had chosen) would go towards your pension. The same would apply to further pay rises: if, in your fifties, you were earning £50,000 more annually than when you took out the pension, you would by then be paying £10,000 each year into a retirement fund. The result was that people taking out a 'Save More Tomorrow' pension would never be poorer because of it – they would just be 'less richer'. To an economist these two states are identical, but to the evolved human brain they are very different.

The idea worked: compared to the control group, twice as many people were willing to participate in the scheme and, of those who did, the average contributions after seven years were roughly double. This can be regarded as alchemy, because it has created a change in behaviour without requiring any material incentive – it simply offers a behaviour that is more congruent with how our brains really work.

No less significant was the success of the UK Government in creating pension-saving by default when they introduced auto-enrolment. As a result, over seven million people now have pensions who did not have them before.

We are a herd species in many ways: we feel comfortable in company and like to buy things in packs. This is not irrational – it is a useful heuristic that helps avoid catastrophe. Antelope might be able to find slightly better grass by escaping their herd and wandering off on their own, but a lone one would need to spend a large proportion of its time looking out for predators rather than grazing; even if the grass is slightly worse with the herd, they are able to safely spend most of their time grazing, because the burden of watching for threats is shared by many pairs of eyes rather than one. Consumers have a similar instinct – we would rather make a suboptimal decision in company than a perfect decision alone. This is also sensible, even if it isn't conventionally 'rational' – a problem is much less worrying when shared.¶

One of our ideas for making pensions more appealing was to exploit this herd mentality: if you sold pensions to groups of people who all knew each other – the members of a sports club, for instance – the likely levels of trust would be much higher.

Here are a few of our other suggestions:

1. Tell people how much the state is giving them. A tax rebate is a strange form of incentive, delivered in a way that renders it almost invisible: the increment is not paid to you, but added to your pension contributions, where it soon passes out of sight. What if you received a text every month from the government, saying, 'We have contributed an extra £400 to your pension this month'?**
2. Restrict how much people can save. In the UK, you can pay a ridiculously high amount into a pension and still receive government tax relief. At first glance this seems logical – the more people save, the better – but it doesn't create a sense that people are missing out if they fail to reach their allowance. It may seem crazy to say that, 'If you want people to save more, allow them to save less',

¶ Imagine again what you would do if you were sitting happily at home after dark and suddenly all the lights went out. If you are anything like me, your first reaction would be to look out the window and see whether the other homes on your street are affected. If they are all plunged in darkness, you heave a sigh of relief: 'Thank God for that – it's just a power cut, and someone else will need to sort it out.' The alternative is far worse: 'Oh shit, it's just me affected – I'll have to sort this out for myself.'

** If I gave someone £400 every month, I'd make quite a lot of noise about it.

but this is the sort of counterintuitive solution that often appears in psychological alchemy.††

3. Make pension contributions flexible. In the modern-day gig economy, where wages may not be constant, it would not be difficult for contributors to be sent a text every month asking if they wished to a) maintain their normal payment, b) increase it, or c) take a break from payments.
4. Slightly decrease with age the size of the tax rebate offered, in order to give a clear incentive for people to start saving sooner.
5. Enable people to draw money from their pension before they retire. It is ridiculous to be paying 25 per cent interest on a credit card while you have £100,000 sitting in a pension account. And if people want to take a year off work to travel, why shouldn't a pension fund this?‡‡

Even if you disagree with some of these suggestions, I think you would probably acknowledge that some combination of them would motivate saving more effectively than the present system does. What is really telling is that, if you assume that economics needs to be objectively 'true', none of them§§ would even be considered.

†† After all, if it worked *and* made sense, someone would have done it already.

‡‡ I can't see any clear reason why it is immoral to take a year off work in your forties, while stopping work altogether at the age of 60 is completely accepted. The pension was designed for a time when most people were dead at 65 and when work often involved gruelling physical toil – we should question whether this is still relevant.

§§ Except perhaps four.

7.5 ALCHEMY LESSON THREE: FIND DIFFERENT EXPRESSIONS FOR THE SAME THING

Imagine a set of four cards is placed on a table – each of them has a number on one side and a colour on the other. The visible faces on the cards are 5, 8, blue and green. Which card (or cards) must you turn over in order to test the proposition that if a card shows an even number on one face, then its opposite face is blue?*

If a card shows an even number on one face, then its opposite face is blue.

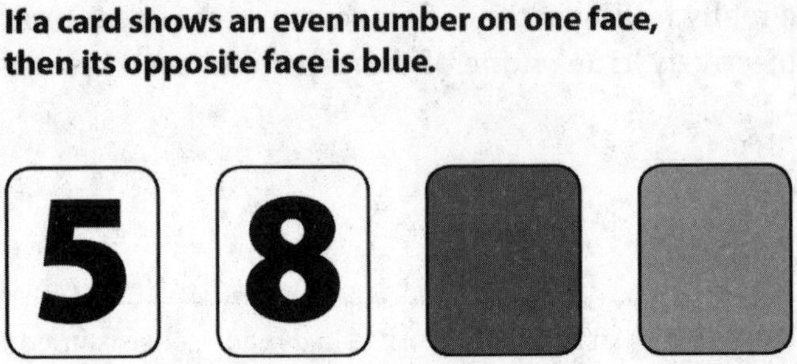

Wason Cards – and how context matters. This test baffled most Princeton students.

....

*   This is called the Wason selection task, originated in 1966. See P. C. Wason, 'Reasoning', in B. M. Foss (ed.), *New Horizons in Psychology* (1966).

A surprising number of bright people get the answer wrong – including, in one test, the majority of Princeton students. Overall, fewer than one in ten people gets this right first time, though no one has any difficulty understanding the problem once the correct answer is explained.

The most common mistake is to assume that you should turn over the blue card; in fact, because the rule says nothing about odd numbers at all, a blue patch accompanied by either an odd or even number would not invalidate the rule – hence there is no need to turn over the blue card. Instead you should turn over the *green* card, since if this were backed by an even number the rule *would* be broken.

As evolutionary psychologists Leda Cosmides and John Tooby observed, if the same problem is framed in the language of social relations rather than in the rarefied language of logic, the success rate is much higher. For instance, imagine instead that the rule is that you must be over 21 to drink alcohol – on one side of the cards is the age of a drinker, and on the other is the drink they have in their hand.

In this presentation of the problem, nearly everybody gets it right: they check the 19-year-old and the age of the person drinking the beer. The can of Coke doesn't matter, and the person over 21 can drink whatever they like. No one has any problem coping with that logic, even though it's the same problem as previously, just framed in a different way.†

†† In 'Cognitive Adaptions for Social Exchange' in J.Barkow *et al.*, *The Adapted Mind* (1992), Tooby and Cosmides propose that the brain contains various evolved modules that are necessary for coping with different processes – when the card problem is presented as a rule-breaking issue, we have no

Yet reframe the same problem in a different way, and any child can do it.

The job of the alchemist is to find out which framing works best. I persuaded my father to pay for TV at the age of 82, simply by reframing the cost. He begrudged paying £17 a month for a satellite television package – it seemed like a waste of money to him. However, when I pointed out that £17 each month worked out to around 50p a day and he already spent £2 each day on newspapers, everything changed. As 50p a day rather than £17 a month,[‡‡] the same cost seemed perfectly reasonable.

difficulty solving it, as part of our brain is optimised for such problems. However, when the same conundrum is presented in the less useful language of pure logic we find it difficult, as we don't have a corresponding module for such abstractions. I am not wholly convinced by this, but it is an interesting idea – and a fascinating experiment.

‡ Granted, a slightly inexact use of mathematics.

7.6 ALCHEMY LESSON FOUR: CREATE GRATUITOUS CHOICES

Provided it is mentally painless, we tend to like choice for its own sake. In the early 1990s, I was working with the recently privatised British Telecom (BT), one of the agency's largest accounts. They had recently modernised the telephone exchanges throughout Britain, and were as a result able to offer customers radical enhancements to their service. For a few pounds a month you could divert calls to another number or subscribe to 'Call Waiting', which let you know if someone else was trying to reach you when you were on the line.

To explain the new services, we sent letters to customers and invited them to subscribe. They were able to opt-in in two ways: either by calling a freephone number, or by ticking a box on a pre-personalised form and returning it in a prepaid envelope – so far, so boring. However, BT had an aversion to allowing people to request the products by post: their argument was that because they were a phone company, we should drive people to use their telephone rather than giving money to the Post Office – they wanted to send out the letters and simply list a phone number as the only mode of response.

To test this, we divided customers into three randomised groups. The first group was offered the choice of responding by phone and post, while the second was only allowed to respond by phone and the third was only able to reply by post. We sent 50,000 letters to each group, and when the responses began to come in, it was soon clear that something strange was going on. The people who had only been offered the chance of responding by phone had a response rate of about 2.9 per cent, and those who only had postal coupons had a response rate of about 5 per cent. But the group who had the choice of responding either by coupon *or* by phone had a response rate of 7.8 per cent – similar to the sum total of the other two. In economic terms, this was bizarre.

People seem to like choice for its own sake.

This is one reason why public services and monopolies, even when they do a good job objectively, are often under-appreciated – it is harder to like something when you haven't chosen it.

It completely mystifies me why most online retailers do not offer you a choice of couriers to deliver your goods. People would vastly prefer this and it would have the additional benefit that they would not wholly blame the retailer if the goods were late or failed to arrive.

7.7 ALCHEMY LESSON FIVE: BE UNPREDICTABLE

Control Tower: 'Maybe we ought to turn on the search lights now?'

Kramer: 'No ... that's just what they'll be expecting us to do.'

Most of business is run according to conventional logic. Finance, operations and logistics all operate through established best practice – there are rules, and you need to have a good reason to break them. But there are other parts of a business that don't work this way, and marketing is one of them: in truth, it's a part of business where there's never best practice, because if you follow a standard orthodoxy your brand will become more like your competitors', thus eroding your advantage. The above joke from *Airplane! (1980)* appears when the air traffic controller is trying to follow protocol, by turning on the lights on the runway for the approaching plane; Kramer, a war veteran, is frightened of being too predictable.* It underlines a serious point.

The marketer's life can be difficult and lonely. Typically, most of a company's management will have the mentality of the air

* Military strategy is in some ways very much like marketing – you can't be conventionally logical as a military strategist, because the enemy will be able to predict what you are going to do.

traffic controller, with a love of the obvious, whereas the marketer needs to be more like Kramer, with a *fear* of the obvious. The two mindsets don't always make for easy bedfellows, and departing from accepted logic can be risky – remember that it is easier to get fired for being illogical than for being unimaginative. Though in many social or complex settings being entirely predictable is hopeless, we tend to fetishise logic.

As Bill Bernbach observed, conventional logic is hopeless in marketing – as you end up in the same place as your competitors.

7.8 ALCHEMY LESSON SIX: DARE TO BE TRIVIAL

The combination of 28 words and a button in the below picture has been called 'the $300m button', and is frequently cited in articles about web design and user experience. It first appeared on an unnamed retail website, which many experts believe to be Best Buy.

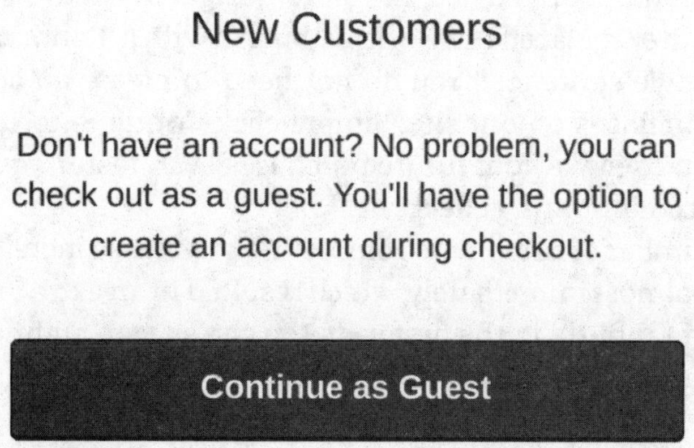
New Customers
Don't have an account? No problem, you can check out as a guest. You'll have the option to create an account during checkout.
Continue as Guest

'The $300m button'. In fact, monumental effects of this kind are surprisingly common in web design. Perhaps one of the first rules of interface design is 'don't try to be logical'.

Jared Spool, the creator of the button, describes the form that customers from the website previously encountered when they came to complete a purchase:

'The form was simple. The fields were *Email Address* and *Password*. The buttons were *Login* and *Register*. The link was '*Forgot Password*'. It was the login form for the site. It's a standard form users encounter all the time. How could they have problems with it? [But] we were wrong about the first-time shoppers. They did mind registering. They resented having to register when they encountered the page. As one shopper told us, "I'm not here to enter into a relationship. I just want to buy something." Some first-time shoppers couldn't remember if it was their first time, becoming frustrated as each common email and password combination failed. We were surprised how much they resisted registering. Without even knowing what was involved in registration, all the users that clicked on the button did so with a sense of despair. Many vocalised how the retailer only wanted their information to pester them with marketing messages they didn't want. Some imagined other nefarious purposes of the obvious attempt to invade privacy.'*

Acting on Spool's advice, the site's designers fixed the problem simply – they replaced the 'Register' button with a 'Continue' button and a single sentence: 'You do not need to create an account to make purchases on our site. Simply click Continue to proceed to checkout. To make your future purchases even faster, you can create an account during checkout.'

The number of customers completing purchases increased by 45 per cent almost immediately, which resulted in an extra $15 million in the first month; in the first year, the site saw an additional $300 million attributable simply to this change.

---

* Jared Spool, in L. Wroblewski, *Web Form Design* (2008). In reality, the site didn't ask for anything during registration that it didn't need to complete the purchase: the customer's name, shipping address, billing address and payment information.

So, people hate registering, and you can increase sales spectacularly by allowing them to bypass registration? Well, it's not quite that straightforward – there's a stranger aspect to this story, which is that most of the site's customers (90 per cent or so) who chose to 'continue as guest' were subsequently happy to register as customers once they had made their purchase – the very people who had baulked at registering *before* completing the purchase were only too happy to leave their details and create an account at the end of the process. This shows that what mattered was not the actions we asked them to perform, but the order in which they were asked to make them.

Typing your address in order to confirm where your new washing machine should be delivered feels like a good use of your time; performing the same task when all you seem to be doing is adding your details to a customer database feels like a waste of your time.

The same thing in a different context can be pleasant or annoying. It's that airport bus all over again.

7.9 ALCHEMY LESSON SEVEN: IN DEFENCE OF TRIVIA

The great copywriter Drayton Bird was once chastised by a friend, who said, 'You advertising people, you go very deeply into the surface of things, don't you?' However, although it was intended as a criticism, I think it should be taken as a compliment.

As any devotee of Sherlock Holmes will tell you, paying attention to trivial things is not necessarily a waste of time, because the most important clues may often seem irrelevant and a lot of life is best understood by observing trivial details. No one complained that Darwin was being trivial in comparing the beaks of finches from one island to another, because his ultimate inferences were so interesting.

The mentality of the physicist or the economist assumes that large effects are only obtained by large inputs. The mind of the alchemist understands that the smallest change in context or meaning can have immense effects on behaviour.

CONCLUSION: ON BEING A LITTLE LESS LOGICAL

No one would doubt that it is possible to have too much randomness, inefficiency and irrationality in life. But the corresponding question, which is never asked, is can you have too *little*? Is logic overrated? I didn't set out in this book to attack economic thinking because it is wrong – I think we should absolutely consider what economic models might reveal. However, it's clear to me that we need to acknowledge that such models can be hopelessly creatively limiting. To put it another way, the problem with logic is that it kills off magic. Or, as Niels Bohr* apparently once told Einstein, 'You are not thinking; you are merely being logical.'

A strictly logical approach to problem-solving gives the reassuring impression that you are solving a problem, even when no such process is possible; consequently the only potential solutions considered are those which have been reached through 'approved' conventional reasoning – often at the expense of better (and cheaper) solutions that involve a greater amount of instinct, imagination or luck.

Remember, if you never do anything differently, you'll reduce your chances of enjoying lucky accidents.

* The Danish physicist, philosopher and Nobel Laureate.

This pseudo-rational approach, with its obsession with following an approved process, excludes counter-intuitive possible solutions and restricts solution-seeking to a small and homogeneous group of people. After all, not even accountants or economists use logic to solve everyday domestic dilemmas, so why do they instinctively reach for calculators and spreadsheets the moment they enter an office? The conventional answer is that we deploy more rigour and structure to our decision-making in business because so much is at stake; but another, less optimistic, explanation is that the limitations of this approach are in fact what makes it appealing – the last thing people want when faced with a problem is a range of creative solutions, with no means of choosing between them other than their subjective judgement. It seems safer to create an artificial model that allows one logical solution and to claim that the decision was driven by 'facts' rather than opinion: remember that what often matters most to those making a decision in business or government is not a successful outcome, but their ability to defend their decision, whatever the outcome may be.

SOLVING PROBLEMS USING RATIONALITY IS LIKE PLAYING GOLF WITH ONLY ONE CLUB

You will improve your thinking a great deal if you try to abandon artificial certainty and learn to think ambiguously about the peculiarities of human psychology. However, as I warned at the

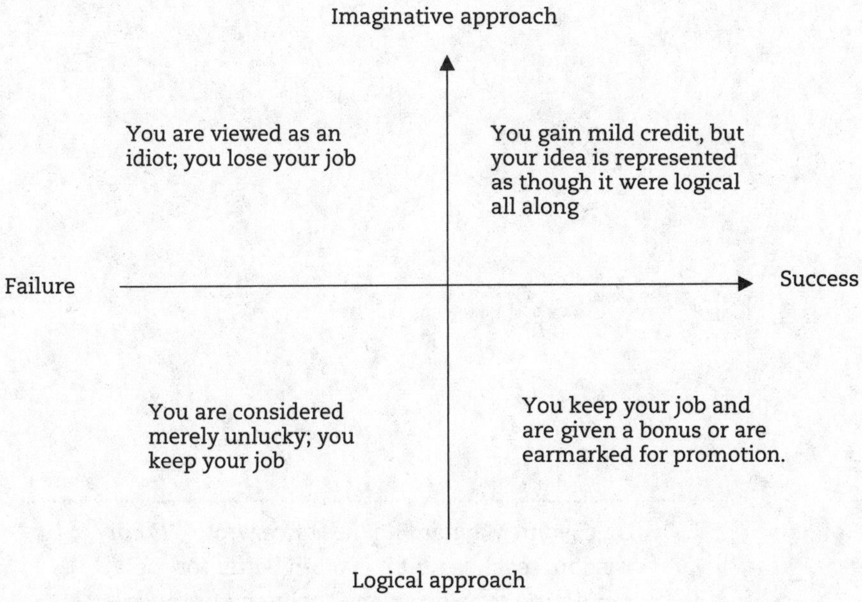
Imaginative approach
You are viewed as an idiot; you lose your job
You gain mild credit, but your idea is represented as though it were logical all along
Failure
Success
You are considered merely unlucky; you keep your job
You keep your job and are given a bonus or are earmarked for promotion.
Logical approach

Why we need to spend more time and energy hunting for butterfly effects.

beginning of this book, this will not necessarily make life easier – it is much easier to be fired for being illogical than for being unimaginative.* The chart opposite describes the consequences of different modes of decision-making, whether things go right or wrong.

Large organisations are not set up to reward creative thinking. As the chart shows, the greatest risks result from an imaginative approach, so it seems safer to act logically. However, it is the job of the alchemist to explore the upper half of this chart occasionally – and managers should give their staff permission and unwavering support when they do so.

* Or, as John Maynard Keynes once wrote, 'Wordly wisdom teaches that it is often better for the reputation to fail conventionally than to succeed unconventionally.'

FINDING THE REAL WHY: WE NEED TO TALK ABOUT UNCONSCIOUS MOTIVATIONS

Our brains present us with a view that is the best-calibrated to improve our evolutionary fitness rather than the most accurate. Being ignorant about your own motivations may pay off in evolutionary terms: it is an inarguable truth that evolution cares about fitness rather than objectivity, and if the ability to present oneself in a good light has certain reproductive advantages, then it will be prioritised. I suspect that we can't overcome these tendencies, and I am not sure that we would even want to, since life would be unrecognisable – and possibly intolerable – without them.

But if we are to understand the power of alchemy, we need better words to describe these motivations, and we must allow ourselves to oppose our natural urge to attach a rational explanation to everything we do. So one of my final tips is that alchemy is not only something you do – it's what you *don't* do.

I am not asking you to read this book and then perform any peculiar feat of intelligence – all that is necessary is to abandon the assumptions that you carry around with you like a comfort blanket every day. The difficult part is to abandon them all at the same time, which carries a risk of social embarrassment. For instance, given the modern open-plan office and our obsession with responding to emails as quickly as possible, it might be embarrassing or even damaging to spend 20 minutes staring blankly into space. However, without this time to disengage, it is harder to practise mental alchemy.*

* Not a single word of this book has been written in my office – just as David Ogilvy did not write a single advertisement in the office ('Too many distractions,' he said). And perhaps 80 per cent of it has been written on days where I had done pretty much nothing the previous day. As John Lennon observed, 'Time spent doing nothing is rarely wasted', yet the modern world seems to do all that it can to destroy the moments where alchemy might flourish.

REBEL AGAINST THE ARITHMOCRACY

My friend, the advertising expert Anthony Tasgal, coined the term 'the arithmocracy' to describe a new class of influential people who believe that their superior level of education qualifies them to make economic and political decisions. It includes economists, politicians of all types, management consultants, think tanks, civil servants and people much like me. I do not believe that these people form a conspiracy and I think most of what they do is intended for the common good. However, they're dangerous because their worship of reason leaves them unable to imagine improvements to life, outside a narrow range of measures. Writing about such people in *The Thing* (1929), G.K. Chesterton explained:

'In the matter of reforming things, as distinct from deforming them, there is one plain and simple principle; a principle which will probably be called a paradox. There exists in such a case a certain institution or law; let us say, for the sake of simplicity, a fence or gate erected across a road. The more modern type of reformer goes gaily up to it and says, "I don't see the use of this; let us clear it away." To which the more intelligent type of reformer will do well to answer: "If you don't see the use of it, I certainly won't let you clear it away. Go away and think. Then, when you can come back and tell me that you do see the use of it, I may allow you to destroy it."'

A huge cast of well-paid people, from management consultants to economic advisors, earn their entire salaries by ripping out 'Chesterton's fences'. Technology companies have partly wrecked the advertising industry and journalism by starving the press of revenue – all under the guise of efficiency. However, they fail to understand that advertising is not really about efficiency – as one expert has put it, 'The part you think is wasted is the part that actually works.' Billions of dollars is now spent on digital advertising because it is assumed to be more efficient – you can target people more accurately and the cost of transmission of each message to a pair of appropriate eyeballs is lower – without it being clear that it is more effective. Procter & Gamble recently claimed to have reduced their digital ad spend by $150 million without noticing any reduction in sales – is it possible that digital advertising is actually strangely ineffectual?

Advertising clearly has a persuasive power that derives from more than just the information it imparts – but where does its power reside, and what makes a television commercial different from a banner? I can think of four things:

1. We know that a television commercial is expensive to make and the airtime is costly to buy.
2. We know that the television commercial is being broadcast to a large number of people, and that those other people are watching the commercial at the same time we are.
3. We know that the advertiser has limited control over who gets to see that message – in other words, that he doesn't choose who he gets to make his promise to.

If the act of advertising generates any of its persuasion through these three mechanisms, it is plausible that digital advertising may *appear* efficient but in reality be surprisingly ineffective.

Remember my argument against Silicon Valley: an automatic door does not replace a doorman. In recent years, advertising could be seen to follow the same pattern:

1. Define advertising as targeted information transmission.
2. Install technology that optimises this narrow function.
3. Declare success, using metrics based on your original definition of function.
4. Capture cost savings for yourself and walk away.

The overly simplistic model of advertising assumes that we ask 'What is the advertisement saying?' rather than 'What does it mean that the advertiser is spending money to promote his wares?', even though we clearly use social intelligence to decode the advertising we see. An example that emphasises the significance of our interpretation of information occurred in eastern Europe under communism; when a product was advertised there, demand often went *down*. This was because under communism anything desirable was in short supply, so people inferred that the government would only promote something that was of such hopelessly crappy quality that people wouldn't be willing to queue for it.

Or imagine you have two products for sale. Product A appears to offer more features than Product B, and is also offered at a lower price. To an economist, the decision is easy: with a higher utility and a lower cost, everyone should buy Product A. However, since a customer is making the decision without perfect knowledge of the two products and their reliability, they might assume that there must be a reason why the price of the ostensibly superior Product A is not higher. The most likely outcome, I suspect, would be for them to buy neither. Whatever economic logic might dictate, the manufacturer of Product A would be better off asking for a slightly higher price than the manufacturer of Product B.

This is not irrationality – it is second-order social intelligence applied to an uncertain world. By using a simple economic model with a narrow view of human motivation, the neo-liberal project has become a threat to the human imagination.

On a trip to Spain before the 2008 global financial crash, I noticed that there were truly monstrous apartment blocks being built mile after mile along the coast.* Construction at the time accounted for an insane 20 per cent of Spanish GDP. I looked at these buildings and asked a simple question. 'Who is going to buy these crappy apartments?' The answer was obvious: no one. Even if the entire population of northern Europe simultaneously decided to decamp to Spain, it was highly unlikely they many of them would choose to live here.

It was soon time for me to fly home; if you flew out of Madrid or Barcelona it was impossible not to notice that both airports were magnificent, but also that they were three times larger than necessary. At London Heathrow or Schiphol in Amsterdam, almost every gate is occupied by a waiting aircraft; here, there were planes at every fifth gate or so. The airports' sheer size spoke to anyone who would listen: if people could easily borrow money for such vanity projects, something had gone badly wrong with the banking sector.

A large part of our brain is designed to consider the messy reality, rather than the neat conceptual theory, yet the use of this part of the brain is generally discouraged. If I had turned up at a meeting about banking with photographs of the substandard apartment buildings on the Spanish coast, I would have been laughed at by economic experts, who would have viewed them as 'purely anecdotal'. Yet as *The Big Short* (2010) by Michael Lewis demonstrated, the people who predicted (and bet on) the failure of the global economy did exactly that – they spoke to estate agents and visited housing developments. Why do we have more faith in a theoretical mathematical model than in what we can see in front of us?

Are we bizarrely cherishing numbers or models over simple observation, because the former look more objective?

* If there had once been a view of the sea, it was now obscured by other, equally horrendous apartment blocks.

ALWAYS REMEMBER TO SCENT THE SOAP

Over the past hundred years, huge improvements in human hygiene have resulted from better levels of sanitation and a growing urge to maintain the appearance of cleanliness, which has brought about a significant change in human behaviour.

When *Downton Abbey* first appeared in 2010, a British newspaper interviewed a nonagenarian aristocrat to ask her whether it faithfully reproduced her memories of the pre-war British country house. 'Well there's one thing it doesn't tell you,' she explained. 'Back then, the servants literally stank.' And in the early twentieth century, when it was proposed to install baths for the undergraduates in one Cambridge college, an elderly fellow was having none of this: 'What do the undergraduates need baths for? The term only lasts eight weeks.'

What had caused this spectacular change in behaviour was complicated, but it was driven as much by unconscious status-seeking as by a conscious effort to improve life expectancy. Soap was sold on its ability to increase your attractiveness more than on its hygienic powers, and while it contained many chemicals that improved hygiene, it is worth remembering that it was also scented to make it attractive – supporting the unconscious promise of the advertising rather than the rational value of the product. The scent was not to make the soap effective, but to make it attractive to consumers.

If we are in denial about unconscious motivation, we forget to scent the soap. If we adopt a narrow view of human motivation, we regard any suggestion of scenting the soap as silly. But, like petals on a flower, it is the apparently pointless thing that makes the system work.

BACK TO THE GALAPAGOS

Because they offer competing choices, consumer markets provide a guide to our unconscious in a way that theories don't. For this reason, I have called consumer capitalism 'the Galapagos Islands for understanding human motivation'; like the beaks of finches, the anomalies are small-but-revealing.

Just as dog breeders and pigeon fanciers understood the principles of natural selection before Darwin codified them, many people involved in selling things have an instinctive grasp of the difference between what people *say* and what they *do*. When he won a MacArthur Foundation fellowship in 1984, Amos Tversky said of his work as a cognitive psychologist, 'What we do is take what is already instinctively known by used-car salesmen and advertising executives, and we examine them in a scientific way.'

We do not have a similar mechanism for politics, or for areas where there is no mechanism for distinguishing unconscious feelings from post-rationalised beliefs. To me, this is the greatest cause for optimism: if we can honestly acknowledge the gulf between our unconscious emotional motivations and our post-rationalisations, many political disagreements may be easier to solve. Again, we simply need to learn to scent the soap.

It has become fashionable to discuss an approach to welfare called Universal Basic Income (UBI). The idea, which has been tested in Finland and a few other places, is to replace welfare programmes with a single minimum income, paid to everybody in the country over a certain age. It would be enough to take care of most people's basic needs; food, heating and housing would be paid for partly by the elimination of other forms of welfare provision but also by higher taxation on higher earners. Whether or not UBI is economically feasible,* it is interesting as a thought experiment – partly because it is surprisingly popular with people on the political right as well as on the left. Milton Friedman supported the idea, as did Richard Nixon. My own grandfather, a man of robust right-wing views, also believed that this is how welfare should work.

People on the political right will normally argue *against* the redistribution of wealth, so what is going on here? Perhaps protests against wealth redistribution are essentially, like most political opinions, merely an attempt to add a rational veneer to an emotional predisposition. People on the right instinctively dislike most welfare programmes, but UBI is paid equally and indiscriminately to all, which means there is no incentive for claimants to exaggerate their own misfortunes in order to benefit. UBI also preserves differential incentives to work: if one man lies in bed all day and his neighbour goes out to a job every morning, the worker will be richer than the layabout in proportion to his effort. Finally, UBI does not allow the ruling political party to bribe its own supporters at the expense of people who don't vote for it.

UBI is an example of a political thought experiment involving 'scenting the soap', in other words lending unconscious emotional appeal to a rational behaviour by changing not what it *is* but how it *feels*. How many more unexpected areas of agreement might we find if we were prepared to experiment with the *presentation* of policies, rather than describing them in narrow functionalist terms? If we spent just 20 per cent of the time we spend preparing economic models on a healthy search for psycho-logical ones, how

* I suspect it isn't.

many more insights might we uncover? Does better psychology, as Robert Trivers wrote, have the potential to uncover and solve some of the deeper roots of our unhappiness?

A few years ago I met Daniel Kahneman for the first time. He was characteristically pessimistic about the prospects of behavioural science to change human decision-making, believing that our biases are just too deeply embedded. However, he was hopeful that people, even if they couldn't see the biases in themselves, might use behavioural science to better understand the behaviour of others. This book has been written in that same spirit. I'm not asking people to completely overhaul all decision-making, to ignore data or to reject facts. But, whether in the bar or the boardroom, I would like just 20 per cent of conversational time to be reserved for the consideration of alternative explanations, acknowledging the possibility that the real 'why' differs from the official 'why', and that our evolved rationality is very different from the economic idea of rationality.

If we could resist the urge to be logical just some of the time, and devote that time instead to the pursuit of alchemy, what might we discover?

Quite a lot of lead, I suspect. But a surprising amount of gold.

ENDNOTES

Page 66 'As the psychologist Jonathan Haidt has shown...', Jonathan Haidt, *The Righteous Mind* (2012).

Page 105 'A recent article in *Harvard Business* Review...', S.K. Johnson, 'If There's Only One Woman in Your Candidate Pool, There's Statistically No Chance She'll Be Hired', *Harvard Business Review* (April 2016).

Page 116 'That's all there is to it.', https://www.farnamstreetblog.com/2009/12/mental-model-scientifc-method.

Page 121 'In the foreword to a WPP annual report...', 'You May Not Know Where You're Going Until You've Got There', *WPP Annual Report* (2014).

Page 161 '... or instruction needed.", Don Norman, *The Design of Everyday Things* (1988).

Page 222 'In the word of Jonathan Haidt...' *The Righteous Mind* (2012).

Page 224 '... offered a possible evolutionary explanation.', Colin Barras, 'Evolution could explain the placebo effect', *New Scientist* (6 September 2012).

Page 240 '... and more by our perception of it', 'The Vodka-Red-Bull Placebo Effect', *Atlantic* (8 June 2017).

Page 295 '... the father of 'Nudge Theory', Richard Thaler' Richard H. Thaler and Cass R. Sunstein, *Nudge: Improving Decisions about Health, Wealth, and Happiness* (2008).

Page 297 '... often outdone by the taste of the latter'.', Lucas Derks and Jaap Hollander, *Essenties van NLP* (1996).

Page 299 '... for leather car seats than for books on tape.", Daniel Kahneman, 'Focusing Illusion', *Edge* (2011).

LIST OF ILLUSTRATIONS

The publisher has made every effort to credit the copyright owners of any material that appears within and will correct any omissions in subsequent editions if notified.

Page 7: Illustration by Greg Stevenson
Page 36: Copyright © Ken Sides
Page 61: Copyright © Benoit Grogan-Avignon, with permission of Shutter Media
Page 106: Recreated by Greg Stevenson, with permission of Stephanie K Johnson
Pages 108-9: Recreated by Greg Stevenson, based on study referred to in Dan Ariely's *Predictably Irrational*
Page 112: Copyright © Country Houses of Kent by Arthur Oswald published by Country Life Ltd.,1933
Page 142: Permission granted by 1stDibs in New York
Page 144: Permission granted by Andrew Heaton
Page 162: Copyright © Shutterstock
Page 183: Illustration by Greg Stevenson
Page 194: Reproduced with permission from Augie
Page 207: Copyright © ICONBIT Mekotron Hoverboard
Page 277: Copyright © worldlifeexpectancy.com
Page 289: Illustration by Greg Stevenson

Page 300: Recreated by Greg Stevenson. From https://sixtysomething.co.uk/compare-breakdown-cover/
Page 304: Copyright © Erwan Mirabeau
Page 305: Image from a study by Greg Borenstein
Page 332: Illustration by Greg Stevenson
Page 334: Illustration by Greg Stevenson